Conceptual
Structural Design

To the memory of Dragoljub & Rada Popovic
and Brian Tyas

Conceptual Structural Design

Bridging the gap between architects and engineers

Olga Popovic Larsen and Andy Tyas

 Thomas Telford

Published by Thomas Telford Publishing, Thomas Telford Ltd, 1 Heron Quay, London E14 4JD.
www.thomastelford.com

Distributors for Thomas Telford books are
USA: ASCE Press, 1801 Alexander Bell Drive, Reston, VA 20191-4400, USA
Japan: Maruzen Co. Ltd, Book Department, 3–10 Nihonbashi 2-chome, Chuo-ku, Tokyo 103
Australia: DA Books and Journals, 648 Whitehorse Road, Mitcham 3132, Victoria

First published 2003

Cover photo: Dutch Expo Pavilion 2000, credit Hans Werlemann. Sketch by Robert Nijsse

A catalogue record for this book is available from the British Library

ISBN: 0 7277 3235 8

Typeset by Gray Publishing, Tunbridge Wells, Kent
Printed and bound in Great Britain by Cromwell Press

Contents

Preface

This book is an essential for both students and young practitioners of architecture and engineering and I wish that it had been around when I was a young engineer. It would have given me the historic background to structural design and architecture over the centuries, which I had to learn from a variety of sources.

The book puts forward a very clear case that collaboration between architect and engineer can only result in better buildings, both architecturally and structurally, and illustrates the huge variety of possible solutions.

The authors put forward two main tenets concerning creativity, firstly: "conceptual structural design and sources of inspiration", and secondly: "art before science". In terms of design they stress the importance of precedents and intuition.

Starting with efficiency of form in nature, this book goes on to provide an historical survey of architecture and structure from the earliest times to the present, including some very interesting examples.

The latter section of the book consists of four fascinating case studies which are very varied, from small to large, in differing materials, and illustrating different thinking. One of the most interesting aspects of these case studies is that they are based on conversations with the architect and engineer for each project and these conversations reveal a true bridging of the gap between architects and engineers.

As a reference, the bibliography gives all the leads that an interested reader would require.

Professor Tony Hunt
October 2003

About the Authors

Olga Popovic Larsen is a lecturer at the University of Sheffield School of Architecture. Her degree in Architecture, Master's Degree in Earthquake Engineering and PhD in Reciprocal Frame Structures enables her to apply an interdisciplinary approach to her teaching. Her teaching in conceptual structural design and architecture seeks to bridge the gap between the two professions. In addition, she is a Course Tutor for the University of Sheffield MEng course in Structural Engineering and Architecture. Her research interests are in the field of advanced structural systems and include reciprocal frame structures, tensegrity structures and membranes. Her research has been widely published in international journal and conference publications.

Andy Tyas is a lecturer in the Department of Civil and Structural Engineering at the University of Sheffield, where he teaches in the areas of structural analysis and structural design. He has close links with the School of Architecture at Sheffield, and has tutored students on a number of interdisciplinary design projects. In addition, he is Course Tutor for the MEng course in Structural Engineering and Architecture which at present is unique in the UK in being accredited by the ICE, IStructE and RIBA. He has published a number of papers at international conferences and in international journals on the teaching of conceptual structural design to both engineering and architecture students. His research interests include the development of computer-aided conceptual structural design programs.

Acknowledgements

The authors would like to express their deep gratitude to a number of people who have helped in different ways to bring this book to completion. The participation and enthusiasm of the designers, Sarah Wigglesworth, Jane Wernick, Dirk Jan Postel, Robert Nijsse, Jacob van Rijs, David Kirkland and Alan Jones, whose projects are discussed in the case studies, was both a vital factor and an inspiration for us. Their help in checking and where necessary amending the text will hopefully produce more accurate and insightful case studies than we would have managed alone. Thank you to Alan Berneau of The Sheffield Anthony Hunt Associates office for providing the information and images about the Don Valley Stadium. Unless otherwise credited, the marvellous hand-drawn sketches are the work of Eleanor Batley, a student in the University of Sheffield School of Architecture. Eleanor spent a great deal of her own time trying to convert our vague suggestions into meaningful images, without which the text would be far less alive. The assistance of Peter Lathey of the University of Sheffield School of Architecture in scanning and preparing images for the book at a particularly busy time of year is gratefully acknowledged. Professor Miklos Ivanyi of the Department of Structural Engineering, Budapest University of Technology and Economics, was particularly helpful in pointing out some of the features of the Szabadszag Bridge, and initiating a thought process on the role of structural correctness versus aesthetic appearance. Without the help of Elisa Gutierrez-Guzman and Konstantinos Sakantamis we would have not been able to collate the text and all the images. We are grateful for your help.

We have dealt with a number of people at Thomas Telford, including Steven Cross, Jeremy Brinton, Alasdair Deas and Mary O'Hara, all of whom have been supportive in helping us produce the final text. Perhaps the biggest thank-you (now that the book is completed!) should go to James Murphy, who originally approached us with the idea of producing a text on the collaboration between architects and engineers.

Without naming them all, we would like to say a big thank-you to friends, colleagues and family for their support and help throughout the writing. Finally, a thank-you to Laura and Jens for their continued encouragement and advice on the text of the book. Last, but not least: a huge thanks to baby Sofia for not waking up crying in the night *too* often!

Chapter 1

Introduction

Many have attempted to define the roles of architects and engineers in the design of buildings. Their roles have changed over time. In the past, at the time of the construction of the great cathedrals, the Master Builder was the person who dealt with all the design issues to do with a building, from the very artistic to the very technical. He was the "architect" and the "engineer" for the project. However, since the Industrial Revolution with the great development in the field of sciences and materials, a clear distinction between the two professions became more evident: the architect came to be in charge of the architectural issues, whereas the engineer was concerned with the more technical issues.

Le Corbusier's view gives a very appropriate explanation of the roles of the architect and engineer.

> "Under the symbolic composition I have placed two clasped hands, the fingers enlaced horizontally, demonstrating the friendly solidarity of both architect and engineer engaged, on the same level, in building the civilization of the machine age".[1]

In the same text Le Corbusier explains how he sees the roles of an architect and an engineer:

> "These [then] are the engineer's responsibilities: the respect of physical laws, the strength of materials, supply, economy considerations, safety, etc. And these [are] the architect's: humanism, creative imagination, love of beauty, and freedom of choice. In my drawing, the engineer's sphere casts a reflection on that of the architect – the reflection of the knowledge of physical laws. Similarly, the architect's understanding of human problems is reflected in the sphere of the engineer."

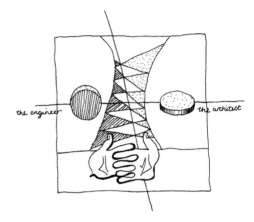

Figure 1-1 The roles of an architect and an engineer, free sketch after Le Corbusier.

If we simplify a project to be a design problem with technical and non-technical design issues, the roles of the architect and engineer can be described as different professionals working within a team, in which the architect comes up with the concept for the project (taking into account contextual, physical, social, political, functional, etc. issues), and the structural engineer deals with technical issues, such as structure, sizes of members, stability, etc. without compromising the architectural concept. And for such a team to work successfully, there must be a mutual respect and appreciation for each other's role, just as on the sketch above.

The development of a structural concept is a critical part of any building design. Getting it "right" can lead to a building which does what

Introduction

the architect envisages, while being financially and technically viable. Getting it wrong can lead to financial or architectural disaster. And herein lies a problem. Engineers are typically educated to think of "design" as being a relatively methodological process for determining the adequacy of a structure or structural member. Architects often think of "design" as being the development of a vision for the appearance and function of a building, incorporating financial, cultural, socio-political and aesthetic factors. So who then is best placed to develop the *structural concept* of a design? Is it the engineer who is not trained to consider non-technical issues in design. Or the architect who is not trained to understand the technical consequences of what may seem to him or her to be trivial decisions about structural form.

The answer, of course, is that the structural concept is (or should be) developed as a collaborative venture. In this, the engineer and the architect must have mutual understanding and respect. The development of a structural concept should be a collaborative process whereby the contrasting requirements of structural necessity, aesthetics and functional utility are synthesized into a workable and impressive whole. Many of the best examples of modern building design where the structure is part of the architecture are the result of a truly combined effort, as we shall describe in this book.

What role does structural efficiency play in defining the concept for a building? There are clearly many other considerations apart from structural efficiency to take into account when deciding on a structural form. However, that brings us to the next question: if structural effi-

ciency is not the only, and in most cases, not the leading factor in the choice of structure, where would the ideas about structural form come from? How do designers decide on what would be the most appropriate structure to make a building stand up?

This book attempts to investigate the above questions. The main focus of the book is conceptual structural design, and the sources of inspiration that have led to developing imaginative structures.

"…at its best [structural engineering design] is an art…it is primarily about the choice of form. The forces on that form and the analysis of its behaviour follow." The above quote[2] by the well-known British engineer Ted Happold,* sums up in a very simple way the standpoint of the authors. Whereas, we acknowledge the importance of technical ability and mathematical analysis, we cannot overstate the importance of creating structural form before complex mathematical analysis begins. Thus, this book addresses conceptual structural design in a non-mathematical way with the main aim of showing how one can arrive at imaginative structural forms using intuition, learning from precedents, understanding of principles and physical modelling.

The phrase 'bridging the gap' works at several levels. At the simplest, the aim of any structural design is to bridge a physical gap. On a more philosophical level, we hope to help bridge the gap between engineers' and architects' understanding of structural form. Finally, the book aims to bridge a gap in the training of both engineers and architects: that is, how to build on their own academic foundations and face the daunting challenge of developing

---

*Founder of the engineering consultancy Buro Happold who have been involved in designing many imaginative structures including the Millennium Dome and the Great Court roof at the British Museum in London.

designs which can be both daring and workable.

The text is presented in two parts. The first is a theoretical part, and it investigates sources of inspiration for developing structural form: learning from natural forms; applying our own intuition; seeking inspiration from precedents; applying understanding of structural principles; and, when the case is too difficult or novel for any of the other methods to work, learning from physical models. These are broadly arranged into the chronological order in which mankind will have studied these sources. It should be noted that this section is most emphatically *not* intended to be a definitive history of development of structural form, nor is it intended to give a full description of all available structural concepts. Our intention is to provide suggestions and hints as to where architects and engineers have found (and continue to find) inspiration for structural form.

The second part is presented through four case studies of built structural forms, in which the process of developing the conceptual structural form is presented through a set of interviews with the design team architect/engineer for each project. The four case studies are different to each other on several levels. They vary in scale: from small, such as the Chelsea Flower Show Pavilion by Sara Wigglesworth Architects with Jane Wernick Associates, to the scale of the Eden Project, by Nicholas Grimshaw Architects

and Anthony Hunt Associates. Somewhere in between, but still very different to each other and also to the first two, are the glass structures by Dirk Jan Postel and Robert Nijsse, and the Dutch Expo 2000 Pavilion by MVRDV Architects and ABT Consulting Engineers. The presented projects also differ on other levels: how the concept was developed, sources of inspiration, function of the building, context, materials used, etc. They all, however, have in common that the design team architect/engineer have been able to develop imaginative structural forms that are in harmony with the architectural concept and, at the same time, are appropriate structural choices from an engineering point of view, being efficient and technically fitting.

The book is aimed at anyone interested in conceptual structural design, architecture and engineering students at all stages of their education and professionals: architects and engineers. The non-mathematical approach will make it accessible to anyone interested in the topic.

Notes

1 Le Corbusier, (1960) August *Science et Vie*.
2 Happold, E., (1984) "The breadth and depth of structural design", in *The Art and Practice of Structural Design – 75th Anniversary International Conference of the Institution of Structural Engineers*, July, pp. 16–22.

Chapter 2

Lessons From Nature: Design Through Evolution

"Never imitate anything but natural forms…"[1]

Natural beauty

From the very beginning of human civilization the fascinating world of Nature has been a continuous source of inspiration to the greatest painters, composers, sculptors, philosophers, poets, designers, architects and engineers. The sound of the running water of rivers, the amazing colours of a rainbow, the regular rhythm of sea waves hitting the steep cliff sides, the captivating smell of roses in blossom, the ingenious organization of a beehive community, the richness of colour in autumn, the dramatic sky during a storm, the snowflakes dispersing the sunlight on a winter's day, are only a few of Nature's wonders that have inspired some of the most beautiful of human creations.

This is not surprising at all because the natural world has reached its present state of development through millions of years of evolution.

The living world of today has gone through many cycles of change and adaptation to reach the level of beauty we see today. In that process, only the most efficient and strongest and most resilient forms of life have survived the challenge of time. In the genes of the living world of today there is a heritage of millions of years of improvement through development.

Aristotle, the classical philosopher, was among the first to write about Nature as a great source of inspiration. He expressed his admiration of the functional beauty found in even the most humble of creatures.

Usually we are inspired by feeling Nature's beauty through smell, colour, shape, form, sound. Often, without looking for the deeper meaning and without questioning Nature's form of expression, intuitively we follow our natural instinct and copy the sounds, colours and shapes that surround us. Undoubtedly, some of the greatest pieces of classical music and paintings have been created in this way.

Natural structure

It is almost certain that the earliest structures employed by prehistoric humans were natural forms. Well before mankind had developed the ability to shape the world around him to his own needs, he would have made use of caves for shelter from the elements, trees for protection from predators or hiding places while hunting, conveniently fallen trees to bridge gaps over ravines or streams.

Figure 2-2 Slender trees in a forest.

Figure 2-1 Landscape in winter.

Figure 2-3 A mountain peak.

Perhaps without even realizing, our ancestors were learning from one of the best teachers of structural form. Nature has its own method of ensuring that its structures are well fitted to their purpose, the most powerful and ruthless method there is; anything which is not strong enough will be destroyed. Thus, erosion in the cave gradually increases the span that the roof must cover until, eventually, the rock is over-stressed and the roof collapses. Or the tree whose roots do not extend far enough into firm soil will one day be put to the test by strong winds.

Over countless millennia, early man gradually changed from an unthinking user of whatever Nature had to offer, to a technological being, able, if only in a limited fashion, to consider his own solutions to the problems of getting by in the world. In this transition mankind must have copied, then later began to adapt examples from the natural world. These lessons are still present today, and cover a huge repository of hard-won "knowledge" which combines structural efficiency with utilitarian practicality in beautiful forms.

Every multiple cell living organism has a structure and must have a structural form. In developing suitable forms for particular tasks, Nature is confronted by many of the conflicting problems which designers have to face: a stronger form may be useful in providing more strength for the organism, but will this mean more weight, and hence, more use of precious raw materials? (And will this lead to a vicious circle, where more weight requires a yet stronger structure to hold it up, requiring yet more weight?) When Nature gets this design balance wrong, the payback is merciless; the hunter that is so heavy that it cannot move quickly enough to catch its prey will not survive long, nor will the hunted which fails to strike a balance between the weight needed for protection and the lightness demanded if it is to be able to flee predators.

For this reason, Nature has an over-riding priority in developing *efficient* structural forms, ones which can support the loads of the organism in a structurally sensible manner, whilst not compromising the organism's operational efficiency. This is one of the key outcomes which a building designer tries to achieve. Isn't the purpose of design to achieve the best outcome for the least outlay?

Nature's combination of beauty and structure: lessons for designers

When we look at a daffodil in blossom we are fascinated by its intense colour, beautiful form and distinct smell. But if we look carefully, and think about it, perhaps it is more fascinating how the daffodil sways in the breeze but does not break and becomes upright as soon as the wind stops. This is more obvious with long grasses that are hinged, which makes them flexible to sway in the wind. It is even more impressive how the stiffening material at the hinges is minimal and is used in the most efficient and effective way. *Nature does not waste*.

There is so much more we can learn from Nature; the lessons go to a far deeper and more subtle level. By attempting to understand the laws of Nature and why natural forms look, feel, sound or smell as they do, we might be able, to a certain degree, to come close to the functional beauty of natural forms (i.e. the beauty achieved through efficiency of material and form). There are fundamental principles in the way Nature develops structural forms, which can be applied to the most modern structures. Many of the finest architects and engineers have taken and applied these lessons to their own buildings.

When considering natural forms, architects have been inspired to try to understand the laws of proportion, curvature, shape and volume, whereas engineers have been analysing the mechanics of materials, stability, motion and considering the underlying principles and efficiency of Nature's creations. Consider the concepts that Nature deals with in producing its forms. It is possible to delete the word "Nature" from Figure 2-4 and replace it with "Design". The concepts dealt with are identical.

Perhaps what makes Nature's creations very special is that beauty of form and efficiency of structure are achieved simultaneously. There is almost without exception a clear logic to the structural principles that somehow in a magical way create aesthetically pleasing forms that we all want to look at and admire, touch and feel in our hands or listen to.

One can find many differences, but also many similarities and analogies between Nature and built form. On a philosophical level Nature has "bridged the gap" (see Figure 2-4 and examples to follow) between structure – logic, efficiency, best use of material properties, functionality – and beauty – proportions, colour, smell, shape, volume – to a level that no man-made built form has achieved to date.

However, on a more pragmatic (and fragmented) level many of "Nature's concepts" have been translated into beautiful and efficient built forms. By understanding the laws of Nature, and applying them to their own creations, humans have been able to produce efficient structures that, to a certain degree, have the shape and proportions that we recognise as "beautiful". When beauty and efficiency have been combined in a structural form, by following the laws

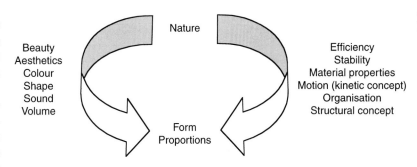

Figure 2-4 The concepts of Nature.

of Nature, our creations, to a certain degree, approach the examples of Nature.

Trees

There is something so natural and so logical, but also so beautiful about how trees look. Look carefully at the structure of a tree: the tree trunk is widest at the base of the tree and the branches are thinner as they get further away from the tree trunk. Trees provide us with a number of lessons on how to produce efficient structures. A tree is, in effect, a large vertical cantilever, which is supported at one end only by its roots. Similarly, the individual branches are smaller scale cantilevers, supported only where they connect to the trunk, or to larger branches. In a cantilever, the stresses are highest at the point of support, and decrease towards the tip. Logically, then, material should be concentrated at the support, and decrease towards the tip, which is exactly what happens in trees.

Designers have known this lesson since the earliest times. As Salvadori has pointed out in his wonderful book on structural engineering,[2] it is not necessary to be a trained structural engineer to appreciate that the cantilever beam shown in

Lessons from nature

Figure 2-5 Structural hierarchy of a tree.

Figure 2-6 "Wrong" cantilever structure.

Figure 2-7 "Correct" cantilever structure.

Figure 2-6 is not efficient. There is something fundamentally unsettling and displeasing, aesthetically "incorrect" perhaps, about this structurally illogical layout. *But* why do we instinctively know that a cantilever shouldn't be like this? Is it because we see a lesson on the correct form every time we see a tree?

The logic of tree structures has been the main inspiration for many structures, some of which are presented later in the text, including: the Eddystone Lighthouse, based on the proportions of an oak tree trunk, the beautiful tree structure roof of Stuttgart Airport, Stansted Airport, and the kinetic tree structure roof design for the Swiss restaurant by Santiago Calatrava.

Jens Larsen

Figure 2-8 Stuttgart Airport, designed by architect Von Gerkan and Marg with engineer Weidleplan Consulting.

Natural stone arches

In many locations around the world, Nature has given us the lesson of how efficient an arch can be at carrying heavy loads over a long span. Pont d'Arc (France), Rainbow Bridge (Arizona, USA) and Landscape Arch (Utah, USA) are among the largest and most famous examples, although smaller versions can be seen wherever rock has been eroded by water. The key to their survival is their shape – rock is strong in compression, weak in tension and, as we shall see (Chapter 4

Jens Larsen

Figure 2-9 Arches carved in stone by rushing water.

on Master Builders) an arch is the perfect shape to transfer loads across a span purely in compression. It should be no surprise that we don't find flat or inverted rock arches – these would generate unsustainable tensile stresses and would collapse.

Eggshells

Eggshells are a beautiful example of a similar principle in miniature. Here, the driving force is to protect the developing young, while using the least amount of material. More material would be both a drain on the mother's body and make it harder for the young finally to break out of the shell. The shell can be thought of as a three-dimensional (3-D) arch, again transmitting the forces efficiently in compression. This 3-D version of an arch is the essence of the domes and shells produced by human designers.

Sea shells

In many cases, because of the shape of an organism, it is not possible to use the arch effect, and bending must be resisted by some other means. This introduces a problem if we want to minimize material. For example, take a flat sheet of paper and flex it. You will feel virtually no resistance to the bending effect. Yet if we introduce a series of folds into the paper (Figure 2-11) we will find that the same volume of paper now has considerable resistance to bending. Nature has beaten us to this discovery. Many sea shells solve the problem of how to minimize material while providing bending strength by having corrugations in the plane of the shell.

Chemical compounds

While not being a direct influence on early humans, molecular arrangements of chemical

Jens Larsen

Figure 2-10 An egg shell providing appropriate strength and protection.

Figure 2-11 Folded paper resists bending better than a flat sheet of paper.

Figure 2-12 Sea shell.

Lessons from nature

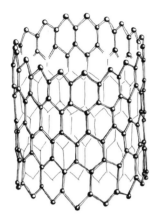

Figure 2-13 The chemical structure of buckminsterfullerene.

compounds suggest that Nature's quest for simplicity, strength and beauty goes to the microscopic level. Primo Levi, the Italian chemist who became a celebrated author, has described a simple and provocative idea of aesthetics derived from the "correctness" of the structural form of carbon molecules (which, by coincidence, happen to be the ones on which all life is based).

"In fact it happens in chemistry as in architecture that 'beautiful' edifices, that is, symmetrical and simple, are also the most sturdy: in short, the same thing happens with molecules as with the cupolas of cathedrals or the arches of bridges. And it is also possible that the explanation is neither remote nor metaphysical: to say 'beautiful' is to say 'desirable', and ever since man built he has wanted to build at the smallest expense and in the most durable fashion, and the aesthetic enjoyment he experiences when contemplating his work comes afterwards. Certainly it has not always been this way: there have been centuries in which 'beauty' was identified with adornment, the superimposed, the frills; but it is probable that they were deviant epochs and that true beauty, in which every century recognises itself, is found in the upright stones, ships' hulls, the blade of an axe, the wing of a plane."[3]

Had he lived to see it, Levi would have undoubtedly seen "beauty" in the structure of buckminsterfullerene, the phenomenally robust artificial allotrope of carbon, which was first produced using nano-technology in the 1990s, and which gives the promise of structural materials of hitherto undreamed-of strength in the future.

Bones

Bones are the components of the skeletal frame which carry the weight of an animal, and, as such, are analogous to the structural framework of beams and columns in modern office blocks, hotels, etc. Nature has an interest in providing the maximum strength of frame for the minimum weight, and has devised an ingenious method for doing so. It appears that bones grow larger and more dense in regions where they are under the largest stress, and wither away when they are not loaded. Thus, a major problem for bed-bound patients, or astronauts in a zero-gravity environment is the loss of bone mass, which isn't being used. Similarly, medical researchers have found that when broken limbs are strengthened by the insertion of metal pins or plates, over time, the new formed bone mass tends to concentrate around the plates in ways which most efficiently mirror the stress concentrations at the interfaces. Nature, in its own way, is trying to produce an efficient

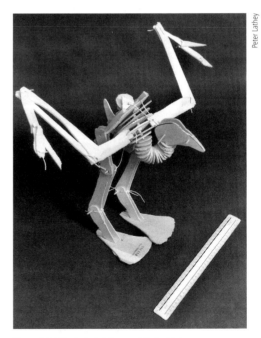

Peter Lathey

Figure 2-14 Student skeleton model of a bird.

structure – components of the frame which are not efficient in helping support it are nothing but a burden, and are slowly eliminated.

The Spanish designer Santiago Calatrava in many of his projects was inspired by Nature and designed bone-like structures; just to name a few: the Kuwait Pavilion in Seville, Spain, the Lusitania Bridge in Merida, Spain and so on.

Skeletal structures

If we try to analyze the load bearing structure of animals, we will notice that many advanced forms of life have a structure comprising an outer skin, which keeps the external environment out and holds the contents in, and a separate, rigid main structure, which is a hidden skeleton. This arrangement gives strength and flexibility; the outer skin can be renewed over time and, if it is locally damaged, can usually be "repaired". Damage to the outer envelope does not usually result in a total loss of structural stability, but damage to the skeleton is more serious and can result in structural failure.

This "skeleton and outer skin" organization with its inbuilt strength and flexibility is an advanced form of natural structure, and one which has developed through evolution over a long period.

On the other hand, in other natural forms, flexibility is not always a vital factor; thus other characteristics predominate. When materials are limited, where a fixed use is envisaged, or where sheer size puts a premium on structural weight, Nature cannot afford the luxury of a separate skin, which adds little to the structural strength. Examples include eggshells (no need to have the flexibility of movement, while the egg-laying creature needs to provide protection for its projected offspring by putting the least possible demand on its own resources) or trees (where size determines that structure and skin must be combined).

The same analogy, essentially, can be applied to building structures. Simpler structures such as traditional houses, churches, etc., are often constructed in a form where the external skin is also the structure. There are, of course, exceptions to this, for example the Indian tent "tepee" has a skeleton and an outer envelope. For the simplest, small-scale (meaning short-span) fixed-use structures, this is often a sensible type of structural form. In structures where a larger free span is required, such as offices, shops, sports halls, etc., a different approach is taken in most cases. The basic structural system is a frame of columns and beams (equivalent to the "animal skeleton"), and the face of the building frequently is used simply to form the external envelope – façade (equivalent to the animal skin), without contributing to the structure of the building. Again, at the largest scales, we may not be able to afford the dead weight of the envelope material which does nothing to help the structure stand up, so we may once again use the skin of the structure as the structural support (e.g. domes, shells).

Lessons from nature: design through evolution

Nature has a vested interest in combining structural efficiency with functionality. It may be argued that the resulting forms, the beating of the wings of a bird in flight, the branches of a tree swaying in the wind, the power and grace of a horse or a big cat in full flight, *define* something that we humans consider to be aesthetically pleasing.

The process of creating natural forms, which we consider to be inherently beautiful, has been a long evolutionary development, that has happened over millions of years. Natural forms have developed by finding ways to adapt to a great range of external factors (climate, availability of food and shelter, etc.) The synthesis of functionality and structural constraints summarizes that multitude of factors.

Random mutations on an already successful theme will usually lead to a non-competitive outcome, but every once in a while will produce an entirely accidental improvement, which, having an advantage over its peers, may go on to become the dominant form. The "failures" will necessarily include those mutations in which either the structural necessity or the functional ideal predominate to such an extent that the other issues are compromised. For example, a chance genetic change that produced a bone which was twice as strong and half the mass of those from a previous generation, but to which tendons and ligaments could not bond, would be useless. Similarly, a bird's wing which could generate twice as much lift, but which was so structurally weak that it would snap when flapped, would be unlikely to be propagated through the generations. Such evolutionary "advances" are almost bound to fail to thrive. Conversely, the "successes" *must* by definition be more profitable and useful combinations of the different requirements of structure and function.

A similar, although less black-and-white, argument applies to design of buildings. All constructions must balance to some extent the requirements of structural necessity with the functional purpose of the building, while as a whole aiming to become an aesthetically pleasing form. However, the analogy with evolution only applies partially, because the design of buildings, apart from structural efficiency and functionality, is influenced by a whole range of political, social, historical or generally "human" factors, also contextual, etc. factors, rather than by the implacable forces of Nature. That is to say, the balance of these design factors is, to a great degree, a matter of choice, and (excluding buildings that collapse, which can be related to natural forms, which became extinct) no building design can be definitively said to be "right" or "wrong" in the way that a thriving or extinct natural creation can.

Imitation of nature

There are two extreme ways in which designers can imitate forms from the natural world around us. In the first, the designer's aim is to mimic or to take inspiration from the *appearance* of natural forms, producing edifices which reflect the beauty of the natural world. In the second, the designer takes inspiration from the *processes* which have shaped Nature's response to the environment.

It ought to be noted from the outset that neither of these two approaches is *guaranteed* to produce good quality architecture. Consideration of *only* the technical lessons from Nature (by understanding the processes) is likely to produce dull and lifeless architecture, because it excludes a whole multitude of "human" and contextual factors. Similarly, thoughtless imitation will lead to foolishly naïve results. It is not surprising, therefore, that among the most criticized buildings are those which seek to copy Nature only in appearance, while missing the underlying "truth" beneath the form. Even John Ruskin, a fervent advocate of nature as a source of inspiration, damned with the faintest of praise the facile imitation of Nature in architecture.

"The fluting of the [Doric] column, which I doubt not was the Greek symbol of the bark of the tree, was imitative in its origin, and feebly resembled many canaliculated organic structures. Beauty is instantly felt in it, but it is beauty of a low order."[4]

Often, in the discussion of structural form, false analogies to Nature are presented. For example, a misconception is often quoted in relation to reinforced concrete, where the analogy is drawn between the concrete surrounding the steel reinforcing bars, and muscle surrounding bone in the human body. On the surface these are similar, but in fact, if we consider the roles of each element (steel, concrete, muscle, bone) in these assemblies (reinforced concrete, muscle and bone), such an analogy is exactly the wrong way around.

The use of muscle and bone to provide bending strength is a wonderful work of natural technology. In the case of the human arm for example, when the hand holds a weight with the forearm horizontal (see Figures 2-15 and 2-16) the forearm cantilevers off the upper arm, and experiences bending. Any structural member (such as the arm) resists bending by generating a tensile and compressive force set up in such a way as to develop a couple acting and thus produce a moment (see Figures 2-15 and 2-16). In the arm, the compressive force is provided by the bone, while the tension is developed in the muscle. In a reinforced concrete beam, it is the steel bars which provide the tensile resistance, while the mass of concrete acts in compression. Despite the superficial similarity between bone, buried deep inside the muscle, and steel bar, buried deep inside the concrete, the two elements perform exactly opposite roles in providing the bending resistance.

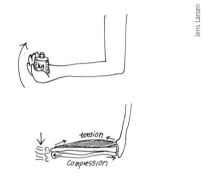

Jens Larsen

Figure 2-16 The bending of an arm – upwards.

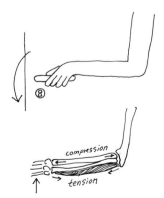

Jens Larsen

Figure 2-15 The bending of an arm – downwards.

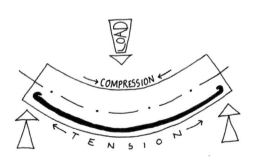

Figure 2-17 A reinforced concrete beam.

Lessons from nature

Despite its lack of obvious immediate resemblance to bone and muscle, a much better analogy of the way in which animal limbs work is the window mullion shown in Figure 2-18. Here, the mullions act as vertical beams to support the glazing panels; since wind can cause pressure to act on the windows in either direction, the mullions must be able to resist bending in both directions – just like a forearm, which must be able to resist forces on the hand acting either upwards or downwards (Figures 2-15 and 2-16). In the arm, this is achieved by muscle placed on either side on the bone – whichever way the arm has to bend; the bone can act in compression while there is a muscle to provide the tension component on the relevant side. In the mullion, note the large, stocky steel bar running down the face of the glazing – that is the compression member. On either side of the glazing panel, attached to the stocky member is a much lighter tension bar.

Whichever way the mullion has to bend, the stocky member acts as the bone, and one or other of the tension bars acts as the muscle. In an example of this kind, whether the result is "beautiful" depends on the proportions of the structural members, the care and attention to design of details and connections, and the eye of the beholder. It is, however, unarguable that such a design faithfully mimics the *processes* of Nature.

Among the most interesting modern examples of architectural lessons from Nature is the recent development of "tree" structures. The purpose of the upper branches of a natural tree is to provide a platform for a large canopy of leaves, which can absorb sunlight and thus provide the tree's sustaining energy. The weight of these upper branches is gradually concentrated into a single, massive and well-founded trunk. Perhaps, if trees sprung into existence fully formed, a more appropriate form in terms of structural efficiency might be a closely spaced grid of mini-trunks, providing both more distributed support and more redundancy against structural damage. But, of course, trees are not created in their finished form. They must grow from seeds, and typically, many saplings will begin to sprout and a considerably smaller number will make it to maturity. Thus, the finished appearance of a tree is intimately connected with the way in which it has grown from a smaller single sapling. Its structural form, the

Andy Tyas

Figure 2-18 Glass mullions at Waterloo Eurostar terminal designed by Nicholas Grimshaw Architects and Anthony Hunt Associates.

massive cantilever trunk taking all the vertical weight and lateral wind load, is a response to this necessity.

Artificial tree structures need to respond to different types of issues. Typically, they are used in buildings with a functional requirement to have widely spaced structural supports, to give a feeling of openness and allow for free and flexible use of the created space. They work best for large open areas, with a sufficiently large floor-to-floor or floor-to-roof height to facilitate the separation of the branches. Typical examples of successful designs include airport departure lounges, meeting halls or transport interchanges. For example, the use of such forms by architect Norman Foster and Ove Arup consulting engineers for the passenger terminal at Stansted Airport, London in the early 1990s allowed them to concentrate the "trunk" columns at around 36 m intervals at ground level, but have the "branch" supports only 12 m apart at roof level. Looking at the form of the tree columns at Stansted, it may be said that both the trunk and branches clearly exhibit their cantilever form, the former gaining bending strength from the squat bundling of four spaced steel tubes, while the latter are combinations of circular steel tube and tension cable.

The analogy with the natural form of trees can be seen on two levels. The superficial appearance is quite obvious, but on a more subtle level the designer is responding to the requirements of the brief to produce an appropriate structural form, just in the same way that a tree is Nature's appropriate response to its own design brief.

It is as well to be aware that the analogy can be pushed too far. Trees and building columns do *not* act in the same way, nor do they have similar loads to carry. Trees are true cantilevers,

Figure 2-19 Stansted Airport designed by architect Norman Foster and Ove Arup consulting engineers.

Figure 2-20 The tree structure at Stansted Airport.

15

and cantilevers are typically relatively flexible structural members. Structurally, trees "give" a little under load. Branches do not remain rigid when the wind blows or when a child swings on them. They flex and gradually build up resistance by doing so. They do not *need* to remain stiff under such loading, and although it is within Nature's power to produce structures which are much stiffer when bent (e.g. bones) the extra complexity, mass and effort required is not justifiable for a tree. Buildings on the other hand, usually *are* required to be much stiffer when loaded. No structure can be infinitely stiff, and all buildings must deform to some extent when loaded, but buildings which move significantly would, at the very least, be unpleasant to be in. At worst, they may be unserviceable due to windows and doors getting damaged as the building flexes, or even dangerous as repeated motion of the load-bearing members can lead to fatigue failure.

Designers of tree structures must bear in mind this requirement for structural stiffness. Making the branches simple cantilevers would naturally lead to them being very heavy members in order to deflect as little as possible under load. A more sophisticated approach, and one which Nature cannot use, is to tie the tips of the branches together at roof level. Thus, branches which would normally deform downwards and outwards, cannot easily do so since they are tied to others which would try to move outwards in an opposite direction. Perhaps the most elegant use of this concept is at Stuttgart Airport departure hall, designed by architect Von Gerkan and Marg with engineer Weidleplan Consulting (see Figure 2-8). The supporting "trees" have many sub-dividing branches, so that the roof structure is supported at only a few metre intervals. Yet, the tying together of the branch tips means that the cantilever members can be relatively slim. The result is a wonderful maze of structure, where structural efficiency, functional utility and aesthetic appeal blend seamlessly together – Nature's lessons learned and applied appropriately, but not dogmatically.

Notes

1   John Ruskin, (1906) *The Seven Lamps of Architecture*, George Allan, London, p. 247.

2   Salvadori, M., (1990) *Why Buildings Stand Up*, W.W. Norton, London.

3   Primo Levi, (1986), *The Periodic Table* (translated from the Italian by Raymond Rosenthal), Abacus Books, London.

4   John Ruskin, (1906) *The Seven Lamps of Architecture*, George Allan, London, p. 188.

Chapter 3

Primitive Structures: Design Through Intuition

Why is a thin piece of cable incapable of being used as a column for a building? Why could we not use bricks and mortar to form a hanging cable? Which is more stable when the wind blows: a tower without foundations, or the same tower with guy ropes?

Of course the questions are simple to the point of being trivial. Yet they illustrate a hugely important fundamental fact. The question of where this "*common sense*" knowledge originally came from is intriguing, and will probably never be answered. While none of us are born with an innate knowledge of structure, *everyone* has some basic understanding of the concepts of structural form that are required to

make buildings stand up. All of us are exposed to structural form throughout our lives, and we pick up lessons empirically from the day we first, as babies, try to place one building block on top of another. We know about structures and how they work through swinging from tree branches as children, sitting on a chair, trying to erect a tent, or putting up a washing line. These activities that are part of our everyday lives provide almost imperceptible lessons on structural behaviour. Being constantly exposed to concepts that "*work*" helps us develop a "*feel*" for what is right and intuitively we are able to tell, often without knowing why, that some structures are better than others. The everyday examples, although simple and obvious, illustrate fundamentals of structures and concepts that are not trivial. Designers employ them to identify quickly and without detailed analysis, what general forms are needed for simple structures, and also in developing more complex structures. The intuitive knowledge is invaluable. It helps us design and build structural forms that somehow we know will work, even though we might not know why!

It should, however, be noted that this intuitive knowledge, developed through experience, could be limiting because if we only rely on it when we face concepts that we have not experienced our instinct could be wrong.

Unconscious absorption of lessons, through intuition and trial and error, is the most likely way that the first pre-technological humans would have developed the familiarity with

Figure 3-1 A tower block needs foundations to provide stability.

structural form. This enabled them to start constructing their environment in a systematic way, as opposed to simply using the shelter that Nature gave them. Man has been a builder since before the dawn of recorded history.

The very first humans, and their ape-like ancestors, almost certainly lived in caves, but as humankind made the transition to a rudimentary form of technological competence, the need for constructing purpose-made shelters arose. The need was generated by the nomadic lifestyle, because early humans would travel in search of food and could not guarantee that suitable caves would always be available. Clearly, the tribes that could construct their own shelters would have a major advantage over those who hadn't made the step from cave dwelling. Although these primitive dwellings were built from natural materials as found (tree branches, leaves, earth, animal bones and skins) using very simple tools or none at all, their construction, in the true sense, marks the very beginnings of artificial architecture. Thus:

"The hut was the basic form of architecture, and it contained major elements of shelter".[1]

Figure 3-2 Neolithic pit dwelling.

Many cultures in their primary stages of evolution lived in pits, huts and similar one-cell dwellings. There are many examples of different types of hut-like dwellings developed by different cultures. They are similar in that they were all constructed in a very simple way using locally available materials, and differ in what the available materials were. These to a great degree determined their form, volume and scale.

Another great similarity of these early structures is that they are all a true form of intuitive design. In these very early stages of human development there were no opportunities for developing any analytical knowledge. It was a time well before any educational system existed. Yet, even today we recognize and admire the early conceptual (architectural/structural) designs developed through trial and error. There is an inherent quality in the designs of hogans, pit dwellings, tepees, yurts and other early structures in that they follow very "correct" structural principles. Their designers learned through trial and error and developed forms that "worked". Some of these principles are so "fundamental" that we apply them daily in contemporary structures.

The primitive (in the sense of "early") structures presented are only some of the examples of indigenous structures developed by early civilisations. They are not the only ones, nor do they present a chronological order. Still, the authors felt that useful lessons about intuitive design of structural form can be learnt from them.

Some of the well-known examples of these early dwellings are the *Neolithic pit dwellings*. They were constructed out of timber beams that meet at the top of the roof, supported by a central pole,[2] as shown in Figure 3-2. The roof was clad with smaller branches and leaves that provided shelter. The main living space was most often in natural, or later, man-made underground holes that provided additional security against other tribes, or animals. Therefore, to create this single cell unit, only a roof was constructed and the retaining walls of the pit were used to provide the vertical enclosure.

When one looks at the structure of the Neolithic pit, it resembles a single pole structure, very similar to an open umbrella. Its main differences are in its craftsmanship, which is rather clumsy and does not always provide a waterproof space under the roof. Still, it is a clear precedent of something a lot more sophisticated that would follow.

One of the reasons for building shelters was to be close to the animals. In early times they were the main source of food. But when there were no animals left to hunt, the dwellings had to be abandoned in order to move to the next destination rich with food. This created the need to develop lightweight and easy to construct structures that could be taken to the next living/hunting destination. The nomads in their permanent search for food, created the first forms of demountable and mobile architecture.

An example of this type of mobile structure was the *Indian tent (tepee)* that was built out of timber poles, which met at the top, were tied together and had a stretched animal skin to provide shelter (Figure 3-3). The tepee was a more sophisticated design than the Neolithic pit dwelling. In most cases it used three or four larger section trees for the main poles, and 15–30 (depending on the size of the tent) smaller secondary poles, arranged to create a conical shape that was tied at the top leaving an opening for ventilation. It is interesting that the skeletal timber structure of the tepee was often tilted to enable better resistance against the prevailing winds. The overlapping animal skins that were

the cladding and created the enclosure, provided additional bracing of the structure. Although it is easy to accept that the Indians knew nothing about wind loads in the analytical sense of the word, one must agree that by tilting the tepees, they created structures that would resist wind loadings most efficiently.

Similar tents, called *tupiqs*, were built and used by Eskimos as their summer residences. They were constructed from slender poles made of pieces of antler or willow wands lashed together. Two pairs of crossed poles formed the basic elements of the *tupiq*, with an arc of poles defining the back of the tent and resting on the fork of the rear crossed supports[3]. The main difference (in structural terms) between the tupiq and the tepee was that the tupiq had two pairs of crossed poles, across which ran a ridge-piece, and the whole frame was covered with a tight fitting membrane of seal-skin or hide (Figure 3-4).

It is interesting that, although some structures emerged in geographically very distant places, the similarity of their structural principles is evident. In that sense, the *black Bedouin tents* emerged almost simultaneously in Northern Africa and parts of Iran and Afghanistan. The Bedouin nomads constructed them when travelling in desert parts to create shelter from the hot sun during the day and protection from the wind and cool weather at night (Figures 3-5 and 3-6).

The main skeletal structure consists of timber poles and barrel-vaults with stretched fabric in between and tensioning cables (ropes) anchored in the ground to provide additional stability against the wind. There were differences in the size and form of the tents and also in the materials used for the fabric, depending on the region where they were constructed. There is evidence that in Iran usually the barrel-vaults

Figure 3-3 Indian tepee.

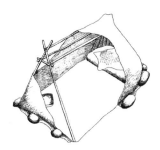

Figure 3-4 Eskimo tent – *tupiq*.

Primitive structures

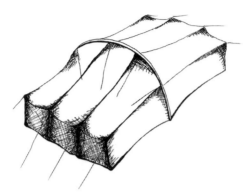

Figure 3-5 Bedouin tent.

were in the longitudinal direction, whereas in other countries they ran in the shorter direction. In some countries there were examples, also, of straight timber poles for the main structure, without any arched forms (Figure 3-6).

The fabric used for the membrane was usually woven of black goat hair or camel hair, but in parts of Syria and Mesopotamia hemp, wool and cotton were also used. Although no contemporary means of structural analysis and design were used, the tent membranes followed similar principles to the ones we use today in the design of fabric structures. In that sense, the Bedouin fabric roofs were composed from strips 70–100 cm wide that were stitched together and reinforced with an additional strip of goat fabric where the fabric was joined. At the support points where the vertical poles touched the fabric, which is a point of very high stresses, small timber insertions were used to reinforce the fabric and protect it against rupture (Figures 3-7 and 3-8).

All these structures, with the exception of the Neolithic pit dwelling, are precedents to contemporary tension structures. One does not need to explain that a fabric cannot span any distance before being tensioned; it is stating the obvious. On the other hand, if a fabric is hung over a structural frame and tensioned by tying

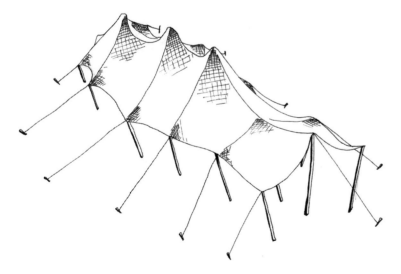

Figure 3-6 Arched form of Bedouin tent.

Figure 3-7 Timber insert at tent ridge.

Figure 3-8 Alternative type of timber insert.

it to the ground, the same fabric works as a membrane structure. It becomes quite strong. This is actually how the primitive tent structures described above work. There is a structural frame which in the case of the tepee consists of several poles tied at the top and anchored at the ground; in the case of the tupiq, in addition there is a ridge pole connecting the two pairs of crossed poles; and in the case of the Bedouin tent the structure is a combination of barrel-vaults and timber poles. A fabric is placed over the skeletal structure. It is then tensioned and tied to the ground.

Many of the basic concepts evident in primitive tent structures have direct comparisons in modern tension membrane structures. An example of a contemporary membrane structure is the Don Valley Stadium roof structure (Figures 3-9–3-11). Built in 1991 for the Sheffield World Student Games, it is among the first fabric roofs in the UK of that scale. It was designed by Sheffield based DBS architects, and the engineers for the project were Anthony Hunt Associates. The cladding is Teflon-coated glass fibre membrane tensioned by high-grade steel cables.

The scale of the roof is very different to the primitive roof structures: Don Valley seats around 10,000 people under the membrane roof, whereas the tepees, tupiqs and Bedouin tents are all single cell dwellings. The materials used are also very different. However, both in the primitive tents and at Don Valley Stadium, the structure consists of a tensioned membrane and the supporting compression structure.

If we go back into history, *yurts* are another form of indigenous demountable architecture. Invented by the Central Asian nomads, they were a very sophisticated design. The yurt was a lightweight type of tent structure that was "designed" and constructed as a prefabricated

Figure 3-9 Don Valley Stadium in Sheffield, DBS architects and engineers Anthony Hunt Associates.

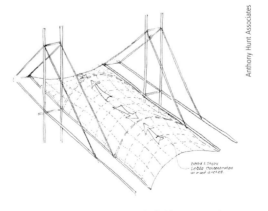

Figure 3-10 Don Valley Stadium in Sheffield, canopy roof.

Primitive structures

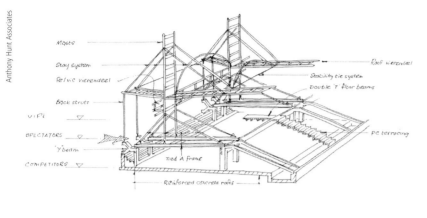

Figure 3-11 Don Valley Stadium in Sheffield, seating and mast arrangement.

When one looks at a yurt and compares it to contemporary built forms, it is obvious that there is logic in this "primitive" design ("primitive" here is used only to describe the time when the yurt first emerged and by no means to describe the structural and constructional principles applied, which are truly ingenious). The nomads, who in search of food moved two to three times a year, were able to take their home with them and re-erect it quickly. The yurt is a genuinely quick and easy-to-build mobile demountable structure, one from which we can learn a lot about structural principles of demountable structures today.

The yurt uses single layer lattice grid shells for the walls and a timber domed structure for the roof. The wall trellises that create the timber grid shell are very efficient lightweight structural forms that are much used in contemporary structures. They use short members that are interwoven and interconnected to create a shell-like structure. All the small members, that in separation do not have a great load bearing capacity, when linked together work as a monolithic structure, which is strong and can resist great loads. This type of structure is a precedent to the contemporary timber grid shells used today.

system and, most probably, was the first true example of this type of structure. It was circular in plan and although quite big (up to 6–7 m in diameter), was a single cell dwelling. Premade trellised wall panels were arranged to enclose the space. They were tied together with leather strings or rope, to provide continuity of the circular wall structure. The walls supported the premade roof structure that was made out of bent timber in the shape of a shallow dome. By laying woollen fabric over the lightweight timber structure and tying it to the timber, weatherproofing was achieved. The process of construction was reversible and when the nomads needed to move, it was easy to take the structure apart and move it to the next destination. The dwelling had to be lightweight so that it could be transported by horses or camels to a new destination. It is interesting to note that the yurts when taken apart could be transported by only one or two horses (or camels) to the next destination.

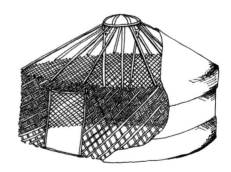

Figure 3-12 Yurt.

Let us look at a contemporary grid shell, like the ones that Frei Otto designed or the recently completed Downland grid shell by Edward Cullian Architects with Buro Happold Engineers that was shortlisted for the Sterling Prize 2002. The small timber elements are interwoven and create a mesh, which in effect is a grid shell. It is a very efficient but also a very beautiful structural form. It is a structure where the engineering and architectural aspects are interwoven and work together in the same way as the structure cleverly shares the load among its members. What is interesting though, is that many years before grid shells had been designed, yurts used trellised wall panels that are true predecessors to contemporary grid shells. And more impor-

tantly they were developed through trial, error and the strong feeling of intuition that we all carry deeply imprinted.

Hogans were used by Navajo Indians, and the oldest type of hogans were constructed from a tripod of forked poles against which others were laid. The pole frame was thickly plastered over with mud. There is also another type of hogan,

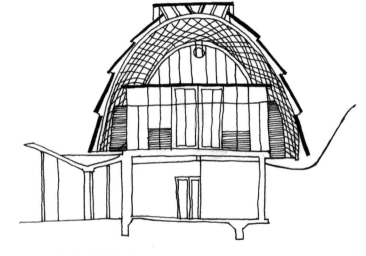

Figure 3-14 Downland grid shell, section.

Figure 3-13 Downland grid shell by Edward Cullian Architects and Buro Happold Engineers, plan.

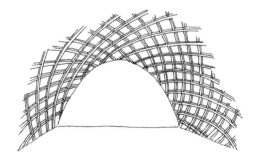

Figure 3-15 Downland grid shell, internal view.

Figure 3-16 Downland grid shell, detail of the grid shell.

23

Primitive structures

Figure 3-17 Hogan dwelling.

with a structure covering a dwelling circular in plan (Figure 3-17). These types of hogans are constructed by laying beams (timber logs or flat long stones) on top of each other and diagonally across the corners of a polygon. The process is repeated, so that the opening becomes smaller and when a dome is formed the structure is covered with mud.

Hogans, unlike yurts, were not designed to be demountable but were permanent homes for their owners. Similarly to the previously described structures (tepees, yurts, neolithic pit dwellings, etc.), they usually created simple one-cell dwellings that enclosed a space circular in plan. However, unlike the previously described structures, the hogans did not have a lightweight membrane (of fabric, animal skin, branches and leaves) like the other dwellings, but were covered with a thick layer of mud. The heavy dome-like structure provided better living conditions and more even temperatures throughout the year. The Navajo Indians who lived in hogans most probably knew nothing about the meaning of the term "thermal capacity". They probably noticed that both animals and humans need a thicker coat (fur) when it is colder, to keep warm. In their designs they started using a new principle, that has since been used many times throughout history and we use it today to calculate wall thickness to suit the thermal performance of our designs. We all notice the even temperatures throughout the year in massive structures. This is especially evident on a hot summer day when we enter a cool cathedral built with thick stone walls!

In a similar way the structure of the hogan follows a very intelligent logic. Because of the heavy enclosure, i.e. dead loads of the thick layer of mud, the structure of the roof is constructed using big sections of timber. The timber structure of a hogan is a lot heavier (stronger) than one of the tepee, but so is the dead load of the mud enclosure of the hogan compared to the lightweight animal skin membrane of the tepee. One sometimes wonders how, in those times long passed, humans who were at their very early stages of development and who had no analytical skills, (before any mathematical calculating knowledge and skills existed), were able to design structures that follow most contemporary structural principles. This was possible, most probably, because conceptual structural design is about understanding structural principles and structural logic and not about calculating! This will be discussed in detail in the chapters that follow.

But if we look to the past, we will notice that our ancestors in their development realized that timber is not very permanent and the use of stone would make their structures more durable. The first ancient masonry structures were a copy of a timber skeletal form. And not surprisingly, they did not work! The Greek temples are a good example of this. The spans of the stone "beams" are extremely short, because the old Greeks found by trial and error that stone, although more durable than timber, unlike timber can only take a negligible amount of bending. The first temples therefore, in a structural sense, look very clumsy and even appear wrong!

Figure 3-18 Greek temple.

Very quickly, however, humans realized that unbound masonry (and even bound masonry) is very weak in bending and that it is impossible to make a beam out of discrete, unconnected elements. Also, they realized that bigger monoliths can span longer distances than smaller ones. Even though they were not able to analyse stresses and strains they were able to work this out.

Most of us intuitively know that arches are stronger than beams and can span further, carrying heavier loads than a beam would, using the same amount of material. The detailed reason why arches work is quite a technical issue, and few people who are not experienced in building theory or practice would be able to give a technical description of it. Yet we can all appreciate that this is the case. Maybe it is because we see domes of cathedrals and bridge masonry arches spanning great distances. But before the cathedrals were built, the first arches must have been developed through trial and error. Mainstone[4] describes how this probably came about by corbelling small masonry blocks over an opening, gradually becoming true arches over centuries or millennia. It was probably through these attempts to build greater and more beautiful structures that men learned how some structural forms work better for certain materials than others. It is how, without analytical skills, they could develop appropriate structural forms that we still admire today.

Another very important issue with all structures is stability. Often structural failure occurs because of loss of stability rather than material failure. If we think about towers, we all somehow know that a tower that is greater at its base will be more stable. The towers we built as children would often fail when the tower was too slender. An excellent example of masonry towers are mountains (Figure 3-19). We all know that their heights can exceed 8000 m. And if we look at a vertical section of a mountain, it is a lot wider at its base than at the top. This is a fundamental lesson on stability, and one that was known by the ancients.

The structures described in this chapter all have in common the fact that they were developed out of necessity, through trial and error, using locally available materials and tools. No structural analysis was done, yet some of these "primitive" structural principles developed many years ago are still in use today. To emphasize this message, we will consider one of the simplest and most intuitive concepts, the stability of columns and struts.

Everyone knows that a slender column, loaded at the top and supported at the base, will tend to topple sideways if the base is not firmly clamped. Many ancient civilizations have left evidence of the awareness of this basic concept in

Jens Larsen

Figure 3-19 A mountain is considerably wider at its base than at its top.

the shape of monumental towers and obelisks, which are invariably well bedded into the supporting ground to provide lateral restraint. Even with this clamping, columns which are heavily loaded can buckle laterally, bending rather than toppling. It is for this reason that tent poles which are either angled together into an "A" frame or whose tops are guyed with tension cables, are more stable than similar poles which are simply thrust into the ground with the tops left unbraced.

Among the largest relatively slender compression members used by the ancients, were the masts of ships (Figures 3-20 and 3-21). The masts would have to support their own weight, the weight of a sail and the weight of sailors climbing up them. In addition, they would have to resist the bending forces generated by the wind on the sails; this had to be transmitted to the rest of the ship's structure and was of course the main motive force for the boat.

Masts would be cut from available trunks of trees, and if they could be made as slender as possible, the boat builder would have a wider and probably more easily accessible selection of timber to choose from. The masts could be firmly fixed into the deck structure, but the tops would have no other supporting structure to restrain them. Consequently, the risk of the mast bending and buckling under the combination of compression and lateral force would be high. A simple solution to this, and one that the ancients used, was to tie the tops of the masts to the deck with angled tension members from either side. Whichever way the top of the mast tended to bend or buckle, it would be restrained by the tension ropes.

As ships became bigger and heavier, the sail and mast system became a much larger struc-

ture. Lashing the top of the mast to the ship's deck might not now be sufficient. Even if the top and bottom of a mast are well restrained, the high compression and bending effects could cause the mast to fail somewhere between these two points. The solution was to introduce more guy ropes, distributed along the length of the mast. This was relatively easy along the length of the ship (as in Figure 3.21), where there was plenty of deck space in which to tie a cable at a reasonable angle. Across the ship there would

Figure 3-20 Ancient masted ship.

Figure 3-21 More recent masted ship.

be less available space; to run a rope from the top of a mast to the side of the boat, the rope would be at a very tight angle and would have to be heavily tensioned to be effective. Instead, the timber cross-spars from which the sails hung could be used to brace the mast laterally, and tension ropes run down from the mast-top to the tips of these spars.

This concept of bracing compression and bending members with tension cables has a direct counterpart today and is increasingly used in buildings to produce self-stabilized members. The aim in such structures is to produce a member where the compression sections are very slender, but where the danger of buckling is eliminated by regularly bracing this member with a grid of tension cables. However the main compression element tries to buckle, it will induce tension resistance in the cables on one side, which will keep it in place. An interesting example is given in Figures 3-22 and 3-23.

A light canopy over a building entrance may be supported by ties running back to the main building structure, but the designer has also to consider the possibility of wind acting *under* the canopy, in which case the ties have to push down against the uplift. A blunt approach to this would be to replace the ties by thicker members, which could provide this compression resistance. In this example however, the designer has taken a more elegant approach, keeping the main load-carrying members relatively slender, and bracing them with a truss of struts and tension cables.

Figure 3-22 A self-stabilized support to an entrance canopy – Singapore.

Figure 3-23 A self-stabilized support to an entrance canopy – Singapore, detail.

Jens Larsen

Figure 3-24 Self-stabilizing structures used for the photovoltaic lighting columns at the Sydney Olympic site.

Considering structural problems on a conceptual, almost intuitive level is often seen as being somehow non-technical, and not sophisticated enough for real design. Nothing could be further from the truth. We all have a fundamental understanding of basic principles of structure, and the application of this *feel* in developing a design concept is something to be fostered rather than ignored. It is particularly vital as designers become more proficient in mathematical analysis. Just as use and a basic understanding of structural form came before the development of structural analysis in the mathematical sense, in a design, in the same way, the almost intuitive development of a concept must come before the analysis. Today's building designers should not forget this lesson.

This approach is increasingly being used also on larger scale buildings and structures, such as the main towers at the Cardiff Millennium Stadium, or the lighting columns at the Sydney Olympic site (Figure 3-24).

Notes

1 Laugier, M.A.,(1977) 1st edn 1755, *An Essay on Architecture*, Hennessay & Ingalls, Los Angeles, p. 3.
2 Crossley, F.H., (1951) *Timber Building in England, from Early Times to the End of the Seventeenth Century*, Batsford, London.
3 Oliver, P. (1987), Dwellings: *The House Across the World*, Phaidon Press, Oxford.
4 Mainstone, R., (1975) *Developments in Structural Form*, Allen Lane, London.

Chapter 4

Master Builders and Beyond: Design From Precedent

Since the earliest days of civilization, building has been a key skill, underpinning civilization. The previous two chapters have shown how "non-scientific" knowledge can act as a foundation for the design and construction of relatively small-scale buildings. When we turn our attention to the largest scale, there is the temptation to think that we can only construct the very largest structures by applying detailed scientific and analytical know-how.

This attitude would have been greeted with disbelief at any point up until the past 150–200 years. Generations and civilizations from the Egyptians, through the Greeks and Romans to Medieval and Renaissance Europeans produced structures on a scale which still has power to impress even today. And for the most part, they did so with little or no use of scientific principles. Their primary method, and the aspect which most clearly set them apart from the builders of pre-technological forms discussed earlier, was a deep knowledge of precedent; what had and had not worked before. This required experience gained from years of working on buildings and learning from their superiors. The key person on such a scheme, the one who had attained an experience and ability above all his peers, was the Master Builder* – the designer *par excellence*.

Achieving the status of Master Builder required a long and dedicated apprenticeship.

Stonemasons would start as boys or young men as assistants to the current Masters, and the ones who proved themselves over the years to be quick and able learners, and able to handle personnel, clients and financiers as well as the material and structure, would eventually take over the mantle when their tutor retired or died. During this long apprenticeship, the future Master would learn what could and could not be done with the materials at his disposal (mainly wood, stone, brick, mortar and, later, iron). He would learn how to shape and cut material in such a way that its natural strength was enhanced rather than compromised. He would learn which structural forms worked and which did not.

The awareness of precedent was a key element in the knowledge of the Master Builder. The Master Builder who studied the work of his predecessors would note the proportions of the structural elements used by previous masters: the relative heights and girths of the columns, the depths and widths of timber beams, and how far they could span between supports. If the buildings, which relied on these supports, had stood for centuries or more, it was clear that the proportions used were "*right*". The Master Builder would learn also, which shape of arch or dome was stable and which was prone to crack and sag; also, how much weight was

---

*The term Master Builder is most often used to describe the head of a construction project in the Middle Ages or Renaissance period, and is associated with the Italian term *capomaestro* – literally "head master". We apply it here more loosely, to cover the designers and builders of large-scale construction work who learned their trade in a more formal way than the builders of primitive forms in pre-technological societies. In practice, this covers the period from the times of the Egyptians and Sumerians until (and in some cases beyond) the Industrial Revolution.

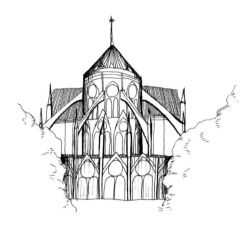

Figure 4-1 Gothic cathedral.

needed to make a buttress stable, and how to sculpt this weight into elegant forms.

In many cases, the proportions of buildings that were of importance (churches, cathedrals or other striking structures) would take on an almost mystical significance, and over the centuries would become the definition of architectural beauty. Indeed, for the Master Builders, the concept was that structural form, strength and stability, and architectural expression were inseparable and complemented each other.

Thus, knowledge of form and proportion was passed down from generation to generation. The secrets of the Master Builders were often jealously guarded, and treated as sacred knowledge; to be knowledgeable of the ancient arts and crafts was akin to a guarantee of work and wealth for a lifetime.

Failures and disasters provided invaluable insight for the Master Builder. Floods, storms, earthquakes all tested structures to their limits, and often beyond, and the inquisitive Master Builder would wish to learn the lessons of why a certain structural form had withstood the catas-

Figure 4-2 Ziggurat.

Figure 4-3 Egyptian pyramid.

trophe, while its neighbour lay in ruins. With no intention to present an historical overview, it is worth mentioning some of the ancient structures from which the Master Builder learnt.

The Master Builders looked at ancient structures and tried to understand the essence of the design principles. The early forms of pyramids were the stepped forms known as ziggurats. These were developed in Mesopotamia for the first time where the Egyptians saw them and copied the shape. Used as burial monuments, in the beginning they were constructed out of materials that did not last (sun-dried mud and bamboo). Experiencing the fragility of the materials and with an intention to build an "eternal afterlife home" the Egyptians started using stone. The geometry, size and shape in a symbolic way reflected the eternity of the structure. Furthermore the form of the pyramid is inherently stable and therefore durable.

Masonry structures were developed in the most efficient structural forms. After many centuries, and after several severe earthquakes, Hagia Sophia, built in the sixth century still impresses with its symmetrical dome structure spanning 31 m, supported by arches that transmit the load to four massive piers, which are buttressed by the rest of the structure.

Jens Larsen

Figure 4-4 Hagia Sofia in Istanbul.

Figure 4-5 Medieval bridge in Skopje.

Figure 4-6 Medieval bridge in Mostar.

Clearly, such an approach to the design of structural form must, by necessity, be conservative. The desire to build higher, span further and construct more quickly and cheaply than the previous generation was hindered by the fact that the understanding of *"correct"* structural form came from the knowledge of what previous generations had been able to construct. Advances did occur, but knowledge and confidence advanced slowly over centuries. Indeed, in the general regression of European culture and civilization, which occurred after the

heights of the Greek and Roman Empires, many of the concepts* and lessons so painfully gained by the greats of antiquity were lost.

Structural forms developed by the Gothic Master Builders

The north-western European medieval Master Builders who produced what was to become known as the "Gothic" style, can be seen as the embodiment of the synthesis of architect and engineer. The Gothic style, originating in France and reaching its heights both there and in Germany, was a radical deviation from the "classical" forms passed down by the Greeks and Romans. Whereas the ancients had looked to the circle and the sphere, with the semi-circular arch and hemispherical dome as the epitome of architectural purity, the forms and proportions of Gothic buildings and structures were governed by other driving forces. The forms of the churches and cathedrals that they built were dominated by the vertical, soaring upwards towards the heavens. Domes were replaced by spires, semi-circular arches by pointed arches; both forms emphasized the vertically upward reach of the new style. The Gothic designers were driven by the need to innovate if they were to achieve the desire of their patrons; to build higher, to span further, to build more economically, and to build more beautifully.

When building in stone, the only way to achieve large spans was to use arched forms (see Figures 4-5 and 4-6). Stone beams would snap under their own weight at fairly short spans, whereas in arches, the force is directed along the member, and the stone's natural compressive strength is utilized. Additionally, arches can be

---

*Rumours abounded that the great structures of the old Empires were the work of gods rather than men, and it would take well over 1000 years before designers would produce anything of similar grandeur.

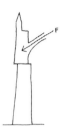

Figure 4-7 Horizontal thrust.

Figure 4-8 The horizontal thrust can cause loss of stability.

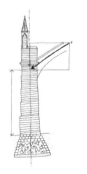

Figure 4-9 Role of the pinnacle in resisting the horizontal thrust.

constructed from a large number of smaller units; laid carefully together to a suitable profile, the component parts of an arch are held together by the compression acting through them. The Master Builders knew by experience that arches and domes gained their strength by a horizontal resistance at their base; remove the horizontal resistance and the arch tends to spread outwards, ultimately leading to collapse. Furthermore, resisting a horizontal push many tens of metres above ground level, for example at the base of a dome, was a difficult challenge, if the mason was to ensure that the supporting walls would be prevented from overturning.

The simplest solution to this, is simply to make the walls supporting the arches massive and heavy; the more self-weight the wall has, the more horizontal thrust it can withstand at the top before toppling. But extra weight did not necessarily mean extra thickness or massive buttresses attached to the walls. Another solution was to build soaring towers and pinnacles above the walls, sometimes towering above the roof itself, adding the necessary weight to the walls and buttresses.

The Master Builders must have been satisfied with a solution which increased the stability of the arches that supported their roofs, while adding the architectural expression of height to their creations, reaching towards the heavens. The resulting form became so ingrained in both technical and aesthetic consciousness, that the concept of a wide-spanning central section of a building flanked by high, heavy towers, became an architectural motif in itself, and has been used far and wide ever since, even in situations where later development of structural form has rendered it unnecessary.

Another solution to the problem of horizontal thrust, common throughout the Gothic period,

suggests a knowledge of the behaviour of arches and the ability to combine desired aesthetic result and structural effectiveness. The horizontal thrust at the support is inversely proportional to the height of the crown of the arch; the higher the arch, the less the thrust. On the other hand, the higher the arch the greater the use of material. The arches and domes often had a pointed rather than a flat, round profile. This pointed profile of the arches of medieval buildings was a major break with the style used in Roman times, and was a probably unconscious approximation to the parabola shape, which gives the minimal material use and maximum strength.

In Roman times the circle was considered by both the pagan and Christian era aesthetes to be the most perfect of forms, and thus the most fitting for use in temples and churches, and by extension the shape of choice for any long-spanning structure – hence, the Roman use of the semi-circular arch in structures from Pont du Gard, in southern France, to the hemispherical domes in the great buildings such as the Pantheon in Rome and Hagia Sophia in Constantinople.

The use of non-circular forms was one of the reasons why Italian architects of the time coined the term "Gothic". They saw themselves as guardians of the ancient truths of architecture and raged against the uneducated and unrefined barbarians. The development of the Gothic style must have been aided by the relative isolation of Dark Ages northern Europe from

Figure 4-10 Pont du Gard.

the heritage of Rome. The first signs of the new architecture, around the eleventh and twelfth centuries, coincided with the Christian crusades to the Middle East. It is highly likely that the beginning of the use of pointed arches was influenced by the Crusaders, noting the predominance of such forms in Muslim architecture, and the resulting experimentation by European builders. Thus, the beginnings of radical new styles such as the Gothic, which developed gradually over several centuries, would have been rooted in lessons from previous builders.

Returning to the issue of the horizontal thrust of an arched or domed structure and how to balance it: while a massive, thick solid stone wall, or high towers were possible and viable solutions, a more elegant solution was to step the buttress away from the wall, and attach stone vaults between the walls and buttresses at key locations (see Figure 4-11). Such "*flying buttresses*" work by effectively making the walls wider in the direction resisting the overturning. Flying buttresses are believed to have first been used as an emergency expedient to prop the walls of a cathedral which bowed outwards alarmingly soon after construction. Nevertheless, their subsequent adoption as a standard form shows a willingness to learn a key structural principle from experience; structural forms which follow the flow of force in a building can be more slender than those which act against the force, since the forces will act *along* the structural members. We *can* design buildings whose forms do not logically follow the flow of force, but there is a penalty to pay. Where force is applied *across* a structural member, such as a horizontal thrust at the top of a slender wall, severe bending stresses are generated. A heavy, thick wall overcomes this difficulty by its own self-weight dominating the horizontal thrust, so that the net flow of force is deflected downwards through the wall itself. The Gothic use of the flying buttress enables the use of thin walls to resist the vertical load from a roof, and delicate, but cleverly aligned buttresses to gently direct the horizontal thrust down to a safe level. It would take many centuries (and one more chapter in this book) before mathematicians could *demonstrate* the truth of the principle of eliminating bending by directing force along a member, yet the stone masons were exhibiting a *practical understanding* of it.

The results were spectacular and revolutionary structural forms. Instead of thick walls, cathedrals were built to hitherto unimaginable heights, with fantastic internal clear spaces, and with but-

Figure 4-11 Section through cathedral, showing the concept of flying buttresses.

tresses which, in the finest examples, appeared to be no more than a filigree of stone lace. The fascinating achievement of the Gothic builders was to work in a medium (stone and masonry), which at large scale can only act in compression, with very little tension and bending strength, and yet produce an architecture which was dominated by the feeling of *space* and *light*. This could only be achieved by a deep understanding of structural behaviour, yet one which was based on empathy with material and form, developed through experience of working with buildings rather than mathematical analysis. As such, their understanding of structure was founded as much in the heart as it was in the brain.

Medieval grillage and reciprocal frame structures

In medieval times most floors in buildings with several levels were built out of timber. When the spans of the structure to be built were short, there was no problem. However, for structures that needed to span distances longer than the available timbers, Master Builders had to look for another solution. They found a clever structur-

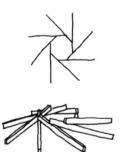

Figure 4-12 Typical medieval floor grillage.

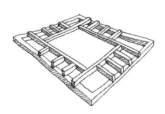

Figure 4-13 Reciprocal frame structure.　Figure 4-14 Honnecourt's planar grillage assembly.

al answer in the form of a planar grillage structure where beams were supporting each other; in structural terms this is very similar to contemporary structures known as reciprocal frames.*

There are many historical examples of this kind. Just to mention a few: the medieval architect, Villard de Honnecourt, who studied the construction of great churches such as Cambrai, Rheims, provides us with information on how to deal with the problem of beams shorter than the span, or as he puts it: *"How to work on a house or tower even if the timbers are too short"*.[1]

Other examples of such planar grillages include the ingenious roof structure of the chapter house at Lincoln, designed by Alexander and built between 1220 and 1235, which is of puzzling complexity but uses timber members that are shorter than the span. Even Leonardo da Vinci (1452–1519) who was one of the greatest of the Renaissance thinkers, as well as the Bolognese painter and architect Sebastiano Serlio, explored the use of structural forms similar to the contemporary reciprocal frame structures.[†] In 1537 Serlio published a prospectus for a treatise on architecture in seven books, and in the fifth book he proposes a planar grillage for a *"...ceiling which is fifteen foot long and as many foot broad with rafters which would be fourteen feet long..."*.[2] He notes that *"the structure would be strong enough"*.[3] In the fourth book, tenth chapter, Serlio made two sketches for door frames which are also planar grillage structures. Serlio's planar grillages are very similar to Honnecourt's solution for spanning long spans with shorter beams.

---

*Reciprocal frames are grillage structures consisting of mutually supporting beams. Usually the term RF is used for grillages consisting of inclined beams, but it also applies to planar arrangements.

†The structural principles of these structures were actually identical, they were simply given different names.

Figure 4-15 Serlio's ceiling.

Figure 4-16 Multiple grillage assembly by Leonardo da Vinci.

What is important is that these structural forms were developed out of necessity, to span a longer distance than the available timber members, and the structural principles used for the structure were developed by intuition, learning through trial and error and also by learning from precedents. In all of these structures there was no detailed structural analysis or detailed "design" as such. The structural concept was developed before any form of complex structural analysis could have been done.

The Master Builder as all-round designer

As building designers increasingly specialized in their main area, all but the occasional genius would lose the detailed knowledge in all fields that the Master Builders possessed.*

The Master Builders, who developed through long practical apprenticeships, were in some cases as close as one can get to the "Universal Man" of Leonardo da Vinci, the person idolised as the aim of human endeavour, the synthesis of scientific understanding and creative ability. In their *combined* understanding of the technical and aesthetic issues involved in the production of structural form, they have few equals today; and even if we can argue that their scientific knowledge was sketchy, their intuition and experience gave them the confidence to produce buildings which still inspire awe today. Many would argue that the combination of scale and beauty of cathedrals is rarely surpassed today, despite our wealth of scientific knowledge.

Such an all-round genius, although one who came from the classical background of Italy rather than the Gothic north Europe, was the Renaissance Master Builder Philipo Brunelleschi, who designed and oversaw the construction of the towering dome above the cathedral of Santa Maria del Fiore in Florence. In his early days, Brunelleschi trained as a painter,[†] sculptor and jewellery-maker. By the age of 24 he was already a renowned artist on the Florentine scene. He entered a competition for the design of the bronze relief decoration of the doors of the Bapistery of San Giovanni in the square dominated by the then half-finished cathedral. Brunelleschi's defeat in this competition, and his subsequent arguments with both judges and winner, led him to leave Florence for Rome – a fortuitous event in architectural history.

In Rome, in the company of his friend and colleague, the sculptor Donatello, Brunelleschi came face to face with the architectural masterpieces of the ancient world. This was the time

---

*Certainly in pre-Industrial Revolution building design, the term "architect" would frequently cover a wide range of skills, including many which we would now think of as the field of the structural engineer.

†He is credited with major innovations in the field of perspective in painting, not perhaps what we would associate with a modern-day structural engineer!

of the Renaissance, and among the artists and the educated there persisted a feeling that the ancient world had taken both aesthetic and technical abilities to their peaks, and that the aim of an educated person was to try to come as close as possible to matching their feats.

Faithful to this ideal, the two friends, making their living as jewellers and goldsmiths, studied many Roman buildings – even earning the soubriquet "The Treasure Hunters" because of their habit of excavating the partially hidden ruins – trying to understand the secrets of the ancient artists and builders. Brunelleschi's biographer[4] notes how he:

"…decided to rediscover the fine and skilled art of building and the harmonious proportions of the ancients and how they might, without defect, be employed with convenience and economy."

To do so, the two friends:

"…made rough drawings of almost all the buildings in Rome…with measurements of the widths and heights…and also the lengths, etc."

One structure in particular appears to have left a deep impression on Brunelleschi. The Pantheon was one of the wonders of the Roman world. The domed roof of this colossal temple spanned a clear distance of 43 m, and rose a similar height above the temple floor. In the early 1400s, when Brunelleschi visited Rome, it was still the longest clear span roof anywhere in the world. It had a hemispherical form, and was made from an early form of concrete known as *pozzolana*. Its cross-section (Figure 4-17) clearly demonstrated that the designers knew that the highest stresses would be experienced near the abutments, for they had thickened the dome around its base. They had also, ingeniously, reduced the weight of the shell where excessive weight would do most damage, towards the centre of the span, by casting jars and *amphorae* into the concrete, to produce voids.

Enriched by his studies, Brunelleschi returned to Florence at an opportune moment. The great cathedral of Santa Maria del Fiore had been under construction for many years, the work of a number of Master Builders. The cathedral was on a colossal scale, with a central drum octagonal in plan and 43 m across. The fact that this was to be spanned by a cupola of an unprecedented scale was no accident. Cathedral building in this era was a form of one-upmanship, and the city of Florence wished to demonstrate its cultural and technical (and thereby, its financial) superiority over Pisa, Siena, Lucca and other rivals. Unfortunately, no-one had been able to devise a workable scheme for construction of the cupola.

The situation at Florence was complicated, in that the huge dimensions of the already-built lower part of the cathedral took up the majority of the space in the piazza, leaving no room for flying buttresses (which most Italian architects of the period would anyway have shunned as barbarian style) or massive abutting structures to resist the enormous horizontal thrust of such a large dome. There are some small abutting structures, but these reach to a height some 15 m or more short of the springing point of the dome, and could therefore offer

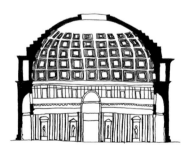

Figure 4-17 The Pantheon in Rome.

little resistance to the thrust. The only structural resistance to this thrust was from the walls of the drum itself, acting as vertical cantilevers. Thus, it was imperative that the horizontal thrust be limited.

The original Master Builder in charge of the design of the cupola was Neri di Fioravanti, famous for designing Florence's Ponte Vecchio. It is clear that Neri was acutely aware of the problem of lateral thrust, and how to resist it. There could be no other reason for his adoption of an obviously pointed dome rather than hemispherical form. As with an arch, the higher the rise of a dome relative to its span, the lower the lateral thrust; such a key technical factor would have had to be of overwhelming importance for Neri and his paymasters to accept such a Gothic form. Almost as radical was Neri's idea for a dome made of two distinct structures, an inner and an outer. The inner structure, the primary load-carrying structure, was an open trelliswork of strong radial stonework members arching from the springing point of the dome to the apex, with lighter circumferential members linking them. The outer structure, designed for appearance and weatherproofing, was a relatively light masonry shell. Neri's design was so convincing that all later Master Builders who were to be involved with the project were required to swear a solemn oath that they would build the dome exactly to the form proposed.

This still left the problem of how to construct the dome. The traditional method of arch or dome building was to erect a temporary scaffold from timber, and support the masonry on this until the structure was completed by the insertion of the keystone at the top, after which the scaffold could be removed. The height and span in Florence were such that this wasn't feasible.* The Florentines were approaching a state of panic. Neri had died without leaving any indication as to how his design was to be realized. Far from demonstrating their superiority, they were in danger of being left with a half-built white elephant![†]

By 1418, Brunelleschi was deeply involved with developing his response to this problem. His previous studies had left him with a sufficiently refined structural feel to comprehend the importance of the double skins and the pointed profile, but it was his breathtakingly daring approach to the problem of construction that has made him one of the most venerated builders of all time. Brunelleschi's logic was simple: the dome could not be constructed *with* falsework, therefore it had to be constructed *without* falsework. This required that, before it was completed (and it would take 16 years for the main structure of the dome to be erected) the part-built structure must be self-supporting. Both skins of Brunelleschi's dome had been developed with exactly this idea in mind and would be built in concentric rings, slowly tapering up towards the apex of the dome. The key was to make each level a self-supporting ring by relying on the compressive arching action in the *horizontal* plane; each piece of masonry would try to fall inwards, but would be restrained by its neighbours which were ever so

Figure 4-18 Florence dome, section.

---

*No scaffold had ever been constructed to such a scale; no timbers, if founded on the floor of the cathedral, could reach the 100 m and more to the top of the dome, or span across the drum if supported from the top of the lower structure.

†It was even suggested that the way to provide a working surface was to fill the cathedral with soil and build the dome off that falsework. To facilitate the tremendous job of then removing the quarter-of-a-million tonnes or so of soil, gold coins could be buried in the mound during construction, and the poor of Florence would willingly come and dig the dirt away in search of treasure!

Figure 4-19 St Peter's in Rome, view of the dome.

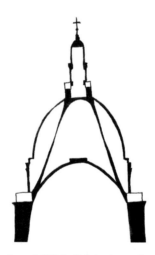

Figure 4-20 St Paul's in London, section through the dome.

exactly to his design, a final indication of his mastery of structural form which is often overlooked. The pointed arch profile of the dome was certainly beneficial in reducing the outward horizontal thrust. But this introduces another, more subtle problem. For a uniformly loaded arch or dome, the ideal shape is a parabolic curve. This shape will eliminate any tendency of the structure to bend, and will ensure that the load is passed back to the supports purely in compression. Significant bending, in stonework or masonry, is disastrous, since it introduces tensile stresses into the material and the tensile strength of such material is very low.

The Roman hemispherical domes and semicircular arches were sufficiently close to this ideal parabolic curve to mean that the bending stresses were minimized. In a pointed Gothic arch, there was usually a large mass of adjacent stonework to help stabilize the arch and provide a parabolic load path. This was not the case in the Florence dome. Brunelleschi's response to this was to add a massive point load of more than 500 tonnes at the apex of the dome by making the lantern, a huge edifice in its own right, an unprecedented weight for a dome to carry. In fact, this makes a pointed dome or arch *the* ideal shape, just as a cable hanging under its own self-weight, with an additional weight at the centre, will automatically adopt a pointed curve profile. This was an intriguing piece of design, in as much as it is not known whether Brunelleschi was fully aware of the structural brilliance of his solution. Yet it is difficult to imagine such an accomplished architect with a great understanding of structural form being

slightly inset from it as they went around the curve. This was relatively easy in the continuous dome of the outer layer, although some thought had to be given to ensuring that each ring was well bonded to the ring below, and thus stable *before* the entire horizontal ring was completed.* In the inner structure, the circumferential arching resistance was provided by the horizontal members joining the main arch ribs. The main ribs, thus stabilized, could be built upward and inwards for a few metres, until their curve took them too far inside the line of the lower stabilized layer and they would tend to fall inwards. At this point, another horizontal ring would be constructed.†

Brunelleschi was to die in 1446, 10 years after the completion of the dome, but before the completion of the colossal lantern which crowns the cathedral. On his death, his will demanded that the lantern be constructed

---

*The answer was to devise a herring-bone bond for the masonry, tying in each layer to the one below rather than simply laying it flat.

†This explains why the horizontal rings are more widely spaced near the springing point, where the dome is more vertical, and the ribs could be built higher without becoming unstable.

unaware of the result of adding a huge mass of masonry to his dome.

Certainly Brunelleschi's creative genius must have influenced both Michaelangelo and Christopher Wren, who in the next two centuries were to design domes on a similar scale at St Peter's in Rome and St Paul's in London. In both cases, multiple skin domes were used, and in both cases, while the outer skin was hemispherical, the inner shell was quite obviously shaped to more efficiently carry the enormous point load of the lantern. Indeed, in the case of St Paul's, the outer "dome" is simply a lightweight timber construction to carry the waterproof envelope of the roof and give the desired hemispherical appearance. The main masonry structural domes are the conical construction bringing the weight of the lantern down directly to the drum, and the lower hemispherical inner dome, which gives the required internal shape.

Modern examples

A major theme emerges from this necessarily brief and selective overview of the Master Builders' art. The importance for building designers of having a deep knowledge of precedents, and knowing what has worked in the past and what has not, cannot be overstated. From before the Master Builders, and certainly since them, a vast body of knowledge on structural form has been developed and refined, aided as we shall see in the next chapter by developments of new materials and scientific analysis of structural form. Awareness of precedent still is a major source of inspiration for modern designers. And, if one is to avoid previous mistakes and appreciate new opportunities, the vast range of potential structural forms and materials makes the knowledge of precedent even more essential.

This application of knowledge of precedent can take many forms, from designs that are reproducing or reinterpreting the *forms* that have previously been used to designs that seek inspiration through the understanding of *methods* used in the past.

A historical line can be traced to Pier Luigi Nervi, who found inspiration in the forms of the past to design his daring modern structures. The Italian master of reinforced and precast concrete applied the Gothic flying buttressing idea to his designs of the concrete aircraft hangar roofs built at Orvieto near Rome in the 1930s, and his Palazzetto dello Sport built in Rome for the 1960 Olympic Games.

In these structures, Nervi gracefully aligned the buttresses to reflect the direction of the line of thrust from the arched or domed roofs, and the mass of the huge buttresses gradually diverted the

Figure 4-21 Orvieto aircraft hangar roof by Nervi.

Figure 4-22 Palazzetto dello Sport, external view.

Olga Popovic Larsen

Figure 4-23 Palazetto dello Sport, internal view.

lateral thrust down towards the ground. Nervi's graceful buttresses had vertical legs to provide additional support for their self-weight.

Nervi's work showed him to be ahead of his time. At the time, it was not possible, even with the most sophisticated structural analysis techniques available, to predict and thus accurately calculate the composite grid shell action of the aircraft hangar structure. The mathematical analysis was oversimplifying the structural behaviour of the hangar structure, and as a result, the calculated sizes for the steel and concrete members were well beyond what was really necessary. In his recent book *Modern Architecture Through Case Studies*, Peter Blundel Jones[5] describes how Nervi solved this problem:

> "Nervi admitted that with the analytical and mathematical techniques available, the loads could not be adequately calculated, for unlike a simple system of beams loading one onto the other, the members would interact in complex ways. He therefore resorted to a physical model of a scale 1:30 in flexible celluloid. Weights were added to simulate loads, and strain gauges showed the distribution of forces. In this way, the operation of the structure could be monitored, and

it turned out cooperation of the ribs made the measured forces much less than those predicted by calculation on the oversimplified basis... Nervi proudly reported after the model tests, that less steel [for the reinforcement] was needed than had been predicted."

It is important to stress here that Nervi did not simply copy structural forms from the past, but by understanding the structural principles, he reinterpreted them and translated them into new materials. He had a great ability for mathematical analysis, but more importantly, he had a physical (often referred to as intuitive) understanding of structural behaviour. And, when mathematical analysis could not provide the right answers, as in the case of the grid shell action of the aircraft hangar at Orvieto, he was able, with the aid of physical models, to demonstrate the true structural behaviour. Physical modelling, as presented later in this book, has proven to be a very powerful conceptual design tool.

Similarly to Nervi, the Spanish engineer and architect Santiago Calatrava also found inspiration in Gothic structures. His use of an inclined buttress at Lyon airport TGV station goes a step further in eliminating the vertical prop, and relying on the bending strength of the reinforced concrete to support the self-weight of the buttress, something that could not have been done by the old Masters working with stone.

It is perhaps not irrelevant that the examples quoted here are from the work of two of the very rare twentieth-century designers who are widely acknowledged to fully embrace both structural engineering and architecture. Both Calatrava* and Nervi† earned the description architect-engineer from contemporaries who could see that they

---

*Santiago Calatrava is among the few who have a formal education both in architecture and structural engineering.

†It is interesting to mention that to date, Pier Luigi Nervi apart from Ove Arup is the only structural engineer who, for his contributions to design, has been awarded gold medals by both the Royal Institute of British Architects and the Institute of Structural Engineers.

were at the forefront of both disciplines. In this, they were among the very select band of designers who represent modern examples of the Master Builder, the fully rounded building designer who can integrate structure and aesthetics.

Today, individual designers (architects and engineers), with few exceptions, do not have the ability to deal with all design issues. In most cases there will be a team of professionals developing the design, unlike the time of Master Builders when one person was the overall designer. It is therefore important that the team architect and engineer must talk to each other and work together from the earliest stages of conceptual design. An architect may have an ingenious design concept, but unless it is developed into a technically viable scheme, the results could be disastrous, at least financially.

A well-known example is the Sydney Opera House. Most will agree that it is one of the best known landmark buildings in the world, with its soaring sail-like structure, and one that still attracts a great number of visitors from all over the world. The architect, Jørn Utzon, whose design for the opera house won the open competition, unfortunately did not work on the design with an engineer to start with. As a result, it was wrongly expected that the beautifully elegant sail-like shapes he had designed would be built using slender reinforced concrete shells, a very efficient structural system widely used in the 1960s.

As the project engineer Ove Arup has written later, the initial form chosen was inappropriate to resist the shells' weight without generating enormous bending stresses. He describes the long and painful struggle to produce a viable structural solution without compromising the beauty of the architect's proposal, noting that the whole problem became a vicious circle; more bending meant thicker concrete shells had to be used, making the shells heavier, meaning that higher bending moments were generated. The final form took several years to develop, and the strains produced by this process led to the architect leaving the project before it was completed, and never visiting the finished building.

Figure 4-24 Sydney Opera House, external view.

Figure 4-25 Sydney Opera House, internal view.

Had Utzon known more about suitable shell forms, he might have been dissuaded from proposing the form that he did, and the world might have been deprived of what is, by common consent, one of its most beautiful buildings.

This example perhaps highlights one of the fundamental points of friction between engineer and architect. The architect, who is unconcerned with technical issues, has a right, perhaps even a *duty* to innovate, experiment and challenge convention. The engineer on the other hand, with responsibility for the adequacy of the structure, must build on certain foundations. If the engineer wishes to develop and extend structural concepts beyond what is known to be possible, it is vital that he or she understands at a fundamental level why similar structural forms in the past did or did not work, and ensures that enough careful thought has gone into the consequences of proposing something different.

Thus any innovator of *structure* should be expected to face the most rigorous criticism before a new form is accepted.

Only by being aware of this condition and accepting it can the design *team* produce architecture which has the power to inspire and amaze. By being their own sternest critics, and their own most eager students, designers of structure can ensure that they innovate from a position of strength and certainty.

Notes

1 Bowie, T., (1959) *The Sketchbook of Villard de Honnecourt*, Indiana University Press, p. 130.
2 Murray, P., (1986), *The Architecture of the Italian Renaissance*, Batsford, London, p. 31.
3 Serlio, S., (1970), *First book of Architecture by Sebastiano Serlio*, first publ. 1619, Benjamin Bloom, New York, p. 57.
4 Manetti, A., (1970) *The Life of Brunelleschi*, University Press, Pennsylvania.
5 Blundel Jones, P., (2002) *Modern Architecture Through Case Studies*, Architectural Press, Oxford.

Chapter 5

Scientific Principles and New Materials: Design Through New Possibilities

Today, the Master Builder is, apart from a few isolated geniuses, an extinct species. The designer who learned all aspects of building technology, aesthetics and construction, was slowly killed off by the radical new opportunities and requirements ushered in by the Industrial Revolution. The radical overturning of accepted processes that this upheaval brought affected all parts of society; they did not miss the building profession. The requirement for faster construction and more economical structures went hand-in-hand with the rapid growth in railway and factory building that the Industrial Revolution spawned. The process was symbiotic. The builders of new factories provided the workshops for industrialists to produce goods on an unprecedented scale, while on the other hand designers now had access to cheap and plentiful building materials. Concrete, bricks, iron and steel began to replace the age-old natural materials, stone and timber.

To take advantage of the new materials, and answer the demands of speed and efficiency placed on them by a newly dynamic society, designers began to require more certainty in their designs. They needed to know with more accuracy how materials performed, what stresses and strains would be induced by the loads that the materials would have to take. This would allow them to make more efficient use of materials, saving weight, time and cost.

Fortunately, a body of scientific knowledge had been slowly building over the preceding centuries. Many of the finest and most famous minds in history applied themselves to the problem of working out theories and formulae to describe structural behaviour. Many radical advances were made in the understanding of how structures and structural elements behave. Isaac Newton had famously laid the foundations for all modern mechanics by his definitive work on equilibrium and forces. The first study of how stresses vary across a bending beam was published by Galileo* in 1638.[1] In 1744, Leonard Euler, possibly one of the world's finest mathematicians, devised a formula to predict the compressive load that a particular column could withstand without buckling.

The result of these efforts and those of other, less celebrated workers, was to develop a large body of theoretical principles which underpin modern analysis of structures. These allow structural analysts to determine with a fair degree of accuracy how their structures will behave under any given load. A large proportion of the education of a modern structural engineer revolves around learning and applying these ideas.

By the early 1800s, technical schools were teaching these theories to a new generation of designers. A split was forming between the engineer, who henceforth would be responsible for the technical design of a building, and the

---

*Even the finest minds can be mistaken. Galileo devised a solution which was later proved to be incorrect. However, his application of scientific reasoning to the problem *was* a major breakthrough, and started a process which would result in the French mathematician Parent devising the correct solution in 1713.

architect, who would consider the social, functional and aesthetic aspects of design. It is a split that has deepened and widened over the last two centuries.

From the mid-1700s, the first tentative uses of mathematical analysis of buildings were being applied, initially retrospectively, to explain such features as cracking in the walls of St Peter's in Rome, then for the analysis of forces in new structures.

Now, here is a vitally important warning: analysis is *not* the same as design. Unfortunately, the word "design" has been misapplied in structural engineering circles to the process of calculating how big a steel beam should be, or how many reinforcing bars should be in a concrete column. The "design" that we are talking about is more concerned with developing a suitable overall *form* for a structure, rather than identifying the proportions of individual elements.

Modern day researchers are working towards developing reliable computer-based methods which can determine the efficient *structural* form to carry a given set of loads between given supports, but such approaches have rarely crossed over from academia into the world of practical design. These computer-based design tools*, if they become commercially available, will undoubtedly help designers to develop structural form by suggesting a variety of possible options. At this stage, however, the state-of-the-art conceptual design software does not take into account issues such as the aesthetics and utility of a structure. Therefore, it seems that, at least for the foreseeable future, conceptual structural design will still require the application of human designers' brainpower.

But let's not swing too far the other way. To ignore structural theory would be as foolish as to assume that these rules are all we need. Indeed, the analytical methods and new materials of the Industrial Revolution were to cause a revolution in structural form. Prior to this time, designers didn't even have a large choice of different structural members because the majority of structures used compression members and simple materials, such as stone, brick or simple concretes. Small bending members were possible using timber, which has some tensile capacity, or deep stone beams (the large depth reducing the bending stresses to levels that stone could resist). To start with, tension chains could be constructed using iron, and were occasionally employed, but were prohibitively expensive for many applications. In addition, they were prone to brittle failure with little or no warning, meaning that they would often have to be greatly over-designed to provide a sufficiently high factor of safety. Architecture on the largest scale was an architecture of mass, solidity and permanence; the accepted aesthetic was one that stated "eternity".

The most important new materials development in the Industrial Revolution were undoubtedly cheap, reliable iron, and later, steel. These materials had tension capacities as high as their compressive strength, unheard of ratios of strength to weight and, in the case of steel, would fail in tension by ductile stretching, giving warning of any overloading.

This opened up a whole new architecture, one based on the expression of lightness allowed by the use of tension members and trusses. Also,

---

*It is important here to emphasize that structural analysis software, as well as presentation software, is readily available. However, the "design tool" to which the authors are referring is one that would be able to come up with a conceptual solution for the structure, i.e. where, what type and shape the structural elements should be.

the use of long spanning flat metal beams (and later, reinforced concrete with steel bars providing the tensile counterpoint to concrete's compressive strength), allowed for bridging much larger distances than had been possible. The Industrial and architectural revolutions had a dynamism which had never been seen before, and which threatened the old certainties.

To someone like John Ruskin, the conservative Victorian, for whom the architectural/aesthetic values of the established styles of architecture were sacred, with proportions and forms developed and proved over the centuries, the explicit use of iron and steel to provide the main support of buildings was to be loathed:

> "Abstractedly there appears no reason why iron should not be used as well as wood; and the time is probably near when a new system of architectural laws will be developed, adapted entirely to metallic construction. But I believe that the tendency of the present…is to limit the idea of architecture to non-metallic work….The rule is, I think, that metals may be used as a *cement*, but not as a *support* [original emphasis]…. But, the moment that the iron in the least degree takes the place of the stone, and acts by its resistance to crushing, and bears superincumbent weight, or if it acts by its own weight as a counterpoise, and so supersedes the use of pinnacles or buttresses in resisting a lateral thrust, or if, in the form of a rod or a girder, it is used to do what wooden beams would have done as well, that instant the building ceases…to be true architecture."[2]

Although Ruskin seems narrow-minded with today's eye, he had a point. The new designs and designers were changing preconceptions of structure and materials.*

The development of mathematical and theoretical advances made in the field of structural mechanics underpinned the use of new design approaches. For the first time, this allowed designers to mathematically *predict*[†] the behaviour of structures in unfamiliar situations, rather than rely entirely on experience and intuition. This gave designers confidence to propose and realize ever more daring forms, with their analytical skills providing a solid foundation for the step into the unknown. Rapidly, the technological achievements of previous generations were surpassed.

An excellent example is the rapid construction of the railways where, by using the newly available metal structures, roofs could be constructed to be far more slender and lightweight compared to the heavy masonry structures used in the cathedrals of previous centuries. However, spanning long distances by using lightweight arch forms brought new challenges to the designers. Masonry arches and domes had been stabilized by their own enormous self-weight, and the effect of temporary wind or snow loading was negligible compared to their great mass. With the new development designers were able to span considerably longer distances, where the self-weight of the structure was *not* hugely greater than the potential imposed loads. Therefore, they had to consider the effect of loads imposed non-uniformly, such as the wind blowing from one direction or another, creating significant bending effects in the structures. The answer to this was to make the arch members from trusses, which had great bending strength and could counter the potential bending stresses created by the wind or snow.

---

*Ironically, Ruskin's criticism of the barbarity of the new architecture in some way mirrors that of the classical architects who had damned the revolutionary northern European cathedral builders of the Middle Ages with the term "Gothic".

†It is perhaps more accurate to say *estimate* since every structural theory is to some extent an approximation of the true behaviour of buildings and members.

Scientific principles and new materials

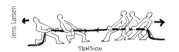

Figure 5-1 Member in tension.

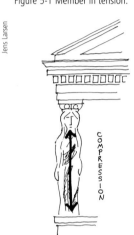

Figure 5-2 Member in compression.

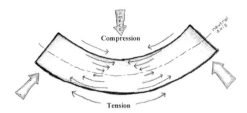

Figure 5-3 Beam in bending.

Trusses made from timber had been used for many centuries in the roofs of large barns, houses and churches, but it was the introduction of high tension capacity metals which really established this structural form. Soon, many major cities had railway stations with roofs that spanned further than even the largest cathedral domes, *and* which, because of the new industrial fabrication processes, had been built in only a few months or years, rather than decades or centuries.

The new approaches led to exciting new possibilities. Prefabrication and standardization of parts meant that enormous and complex structures could be built much quicker than had previously been the case.*

To realize why the new architecture, which in the true sense was an "architecture of light",[†] was so radically different from the old, we need to consider the relative behaviour of members under different forms of loading. The majority of structural members predominantly experience either or both of the two main types of structural effect: direct stress or bending stress.[‡] Direct stress occurs when a member is pre-dominantly loaded along its length, and the whole cross-section is subjected to the same kind of stress. It can take the form of either tension (such members are usually called "ties") or compression ("struts" or "columns"). Bending occurs when a member is loaded *across* its length and results in a combination of both tensile and compressive stresses within the member. These members are called "beams".

If we approach design solely from structural principles, we can define a broad principle which is: "*members loaded in direct tension are generally more efficient[§] than ones loaded in direct compression, which in turn are generally more efficient than ones experiencing bending*". The reason why compression is generally less efficient than tension is that, in compression failure is usually by sideways buckling of the column or strut well before the capacity of the material has been reached, whereas in tension, members tend to fail by exceeding the capacity of the intermolecular bonds.

Tension members	Most efficient
Compression members	Less efficient
Bending members	Least efficient

This suggests that we should try to eliminate bending members wherever possible, and encourage the use of tension members.

Of course, the real world is not so simple. If you look around you will see beams in most structures; supporting the floors of multistorey

---

*The first prefabricated structure on a grand scale was the Crystal Palace designed by Joseph Paxton for the Great Exhibition that was held in London in 1851. This was a building on a colossal scale, but one in which lightness and elegance could tie in naturally with efficiency of design, fabrication and erection.

†Here the expression signifies both a new architecture constructed with lightweight structures, and also one that had spaces that were light (had plenty of sunlight).

‡There is one other type of stress, shear, which is of great importance in the analysis of particular structural elements, but is of lesser importance in the design of the entire structural form, and will be ignored here.

§By "efficient", we mean that a member can carry a given load with the minimum use of material.

buildings, in short road and rail bridges, even the seat of your chair is a bending member, carrying your weight between the supports of the chair legs. In each of these cases, it would be *possible* to replace the beam by a more structurally efficient system of struts or ties but it would not always be desirable.

If, for example, a structural member is to be loaded with 10 tonnes, and the span is 10 m, a solid square mild steel member would need to be around 180 mm square to carry the load in bending (Figure 5-4), 70 mm square to carry the load in compression (Figure 5-5) (assuming some bracing was present to prevent the apex from moving sideways) and only 16 mm square to take the load in tension (this is assuming ideal theoretical elastic behaviour and 45° angles for the members) (Figure 5-6).

But Figure 5-4 also shows the great attraction of beams: they can be essentially flat and still carry loads across a gap. Bridges, floors or chair seats would not be the easiest things to use if they were peaked or troughed like the more efficient compression or tension members in Figures 5-5 and 5-6. In addition, cost and ease of construction are important factors when deciding on the type of structure. While the beam in Figure 5-4 is heavy, it is relatively simple to fabricate and position. Both the compression and tension solutions (Figures 5-5 and 5-6) require some form of additional connection at the centre and/or supports, and both could pose tricky construction problems. Often we will accept the extra material that a beam requires, because the reduction in fabrication and construction costs will more than compensate.

There is a way of combining the efficiency of tension and compression with the utility of beams by using trussed bending members discussed further in the text, where instead of

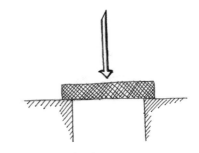

Figure 5-4 Beam in bending.

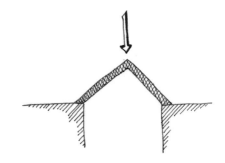

Figure 5-5 Strut in compression.

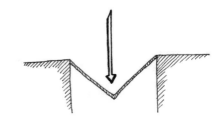

Figure 5-6 Tie in tension.

resisting bending stresses *within* the beam, discrete tension and compression members are trussed together. The big advantage of such a trussed system is that the distance between the centres of tension and compression is typically much larger than in a beam, and so the moment that it can resist, for the same amount of material, is much higher.

Scientific principles and new materials

The availability of new materials, analytical ability and the ability to design structural members* which *can* take large bending stresses, makes it possible to develop *some* structural form to support any building form we envisage. If anything, because of the wider range of possible structural forms, there is even more onus on modern designers to think carefully of a suitable structural system at an early stage in the design process if an elegant and efficient solution is to be found. In terms of this, it becomes more important than ever that the team architect and engineer start working together in developing the structural form as early as possible.

The skill of the modern structural engineer lies in being able to judge when the structural inefficiencies of bending are more than compensated for by the advantage of simplicity of a bending member, and in being prepared to find elegant ways to eliminate bending where it is not absolutely required, or to brace compression members to increase their resistance to buckling. It is important to state at this point that a good working team (engineer/architect) will discuss the efficiency, but will agree on the structural form having in mind all considerations (architectural concept, utility, buildability, context, etc.) and not solely structural efficiency.

The consideration of technical aspects is commonly thought of as being the job of the engineer at detailed design stage. However, awareness by both engineer and architect of the factors involved when determining the overall structural form will aid a holistic approach and one that can often be invaluable in producing both aesthetically pleasing and sensible schemes.

The following example illustrates how designers can apply very different thinking to structures that are superficially similar, but on a different scale. Let us examine how the structure of a small open-fronted bus shelter might work. The major design requirements are that there should be a roof over the heads of the waiting passengers, and an open front to allow easy access to the pavement and overview of the road. Supporting walls or frame can be present at the back and sides of the shelter. We will assume that the shelter is to be 3 m long, 1 m wide and 2.5 m high. The simplest structural form for this shelter would be a cantilever roof, supported by a back wall (Figure 5-7).

The walls and roof could be made from reinforced concrete slabs, or the structural form could be a steel frame with light plastic sheeting between the steel columns and beams. The

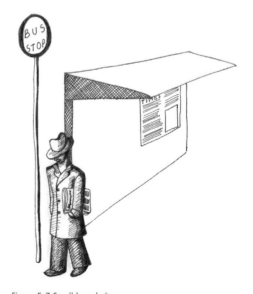

Figure 5-7 Small bus shelter.

*Unlike in medieval times when most structures were compression structures.

roof members would be cantilever bending members, while the back wall or columns would have to resist both the bending from the root of this cantilever and the vertical load in compression. The small scale of this structure means that the bending stresses developed in these members would be relatively small, and could easily be handled by these materials.

Perhaps in acknowledgement of the distribution of bending stresses in a cantilever (natural forms – tree branches), the designer might taper the roof towards the tip, and provide a strong back wall.

As far as the structural behaviour is concerned, the designer here has three main considerations:

- To make the roof structure strong and stiff enough to withstand the cantilever bending produced by loads such as, for example, self-weight, snow and wind (both up and down).*
- To ensure that the back wall (or back columns) is strong and stiff enough to carry the vertical load from the roof plus the bending from the roof cantilever.
- To ensure that the entire assemblage remains stable and does not sustain unacceptable deformations under the above loads.

In architectural terms the problem becomes more complex. The shape/form of the structure should respond to the site and to the surroundings. The response to the site should be the most appropriate[†] and in line with the overall architectural philosophy of the design. In addition, the choice of structural form and material will be influenced by factors such as: whether the shelter is part of a greater development, whether it has a symbolic significance in the whole scheme, how, when and by whom the shelter would be used, etc. Structural efficiency is only one of the considerations in the design of structural form. It is an important one, but a successful design will only result from a holistic approach to the whole multitude of factors involved.

If we go back to the technical–engineering considerations, the three main considerations apply equally if we scale-up the problem and look at the design of a cantilever roof over a sports stadium grandstand. In this case, typical dimensions might be a length of 100 m, a height of 25 m and a width of 30 m, and the increase in size of the problem means that many engineers would take a significantly different approach to the most suitable structural form.

One of the principal forces on the structure is the self-weight of the roof itself, so reducing this load becomes a priority as the structure increases in size. The roof still has to act primarily as a cantilever, but now the convenience of using a simple beam is outweighed by the weight penalty. More self-weight adds to the bending stresses, which demands bigger beams to resist these stresses, resulting in higher self-weight etc. One way to reduce the severity of the bending might be to extend the back wall columns above the height of the roof, and run a cable from the top of the columns to some point along the roof beams, thus providing additional support for the roof (Figure 5-8). An extension of this idea is to replace the roof beam with a lattice truss (Figure 5-9). Here, the bend-

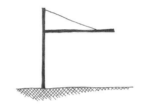

Figure 5-8 We start with a cantilever beam.

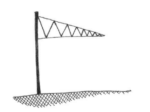

Figure 5-9 The cantilever becomes a truss.

*Even perhaps the weight of overexuberant testers who may wish to climb onto the roof or swing from the tip.

†It could contrast with its surroundings or it could follow the architectural language, accepting and reinterpreting the most relevant proportions of elements, forms and details of surrounding structures.

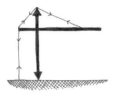

Figure 5-10 Tying the tension force into the ground.

Figure 5-11 The cantilever truss is continuous down to the ground.

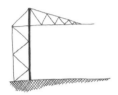

Figure 5-12 Taking advantage of the heavy structure.

Figure 5-13 Tying the trussed roof to the heavy structure.

ing stresses in the roof beam are converted into axial tension and compression in the members of the truss, and large savings can be made in the self-weight of the structure (the penalty being that considerably more time and effort is taken up in detailing, fabricating and connecting the individual members of the truss).

Other than in reducing the self-weight of the roof, this has not helped the problems of bending in the back columns, or overall overturning of the structure. One way to counteract this is to continue the tie, or the truss, all the way down to ground level (Figures 5-10 and 5-11). Now, the bending component of the load in the back columns has been converted into axial tension and compression in the column and tie (Figure 5-10) or the truss members (Figure 5-11). Again, this elimination of beams carrying bending load will radically reduce the size and weight of the structural members.

The overturning effect is now resisted by a moment generated by a couple in the column and tie or the chords of the back truss. Since the tie will tend to pull away from the ground, a heavy mass concrete foundation block, or tension piles, will be necessary to anchor the tie to the ground. Alternatively, the engineer might take advantage of the presence of the heavy structure supporting the terrace under the roof to act as a stable support into which to tie the roof* (Figures 5-12 and 5-13).

Thus, what was essentially the same problem on a different scale, has resulted in a significantly different structural form. The key to this has been the transition from predominantly bending members at small scale, to predominantly axial force members at large scale. This is a gen-

Figure 5-14 Manchester United Football Stadium.

Figure 5-15 The trussed roof is tied to the seating structure at Manchester United.

eral principle, which applies throughout the design of structural form. For example, the world's longest beam bridges, mainly acting in bending, have spans of up to 300 m. Truss bridges, with the bending primarily converted to axial tension and compression, can span up to 500 m, as can arches, where the bending is mainly converted to axial compression. For the very longest spans, currently up to 2000 m, we must use the efficiency of direct tension as in the main cables of a suspension bridge.

In the above presented examples, the designers were able to size the members very easily and quickly because of the advances in structural mechanics. This, together with the great development of software for structural analysis, makes it possible to easily predict the behaviour of a structural configuration.

---

*In this description we have only considered structural issues. In the context of a real design problem the architectural issues will have a great influence on determining the final form, but for clarity and simplicity of presentation they have been omitted.

But if we go back in time, when these advances were happening, designers of buildings began to see the possibilities that using far lighter construction could bring in terms of building higher structures.

Strong, lightweight, long-spanning beams, and perhaps even more importantly, strong columns meant that buildings could be constructed on a much higher scale. Previously, all high buildings were constructed using load-bearing brick or stone walls. These had a very high self-weight, and the walls at the lower levels had to be very thick simply to carry the weight of walls above them. The advent of iron and steel columns allowed an entirely different type of structural form, where the load in the building was directed through discrete columns, rather than continuous walls. The real structure holding up a building therefore became a skeleton of beams and columns; designers could envelope this in a non-load-bearing façade of their choice, first brick and stone, later concrete and glass. On one level, this opening up of the structure followed the age-old desire to use columns instead of walls (e.g. Greek temples, Gothic cathedrals), but the new materials and new certainties of structural performance took the use of columns to a hitherto undreamed of scale. It is interesting to note that the first true skeletal high-rise forms were developed for the reconstruction of Chicago following a devastating fire in 1871, which destroyed almost 20,000 buildings. Thereafter, the growth of high buildings exploded, with structures soon reaching a third of a kilometre and more into the sky and producing the defining image of the twentieth century city.

While the skyscrapers of the USA are perhaps the most potent symbol of this new architecture, the Eiffel Tower in Paris is perhaps the most informative emblem of this newly confident era

of scientific design. Constructed in 1889 for the one hundredth anniversary of the French Revolution, the genius behind the tower was French engineer, Gustave Eiffel. Eiffel had built a reputation and a successful company on the use of iron and steel in a succession of breathtakingly elegant and structurally (and aesthetically) revolutionary bridges, such as the Pont Garabit in Auvergne and the Duoro Bridge in Oporto, Portugal. He was also the designer of the steel truss frame which forms the hidden internal structure of New York's Statue of Liberty.

Eiffel was a ground-breaker in more than the use of the new material. He also developed structural forms suitable for their properties. While many designers were probing the new *structural* possibilities of the new materials and analytical processes, they often clung to the certainties of established *architectural* forms, or at least compromised between the two. A classic example of this approach is the Sydney Harbour Bridge built in 1932, where the two massive stone towers have no structural function, apart from giving the *expected* visual impression of solidity and strength. The newly adopted material, steel, was used to create the arched trusses, which form the load-bearing structure for the bridge, would probably not have been sufficiently reassuring for the general public at

Figure 5-16 The Eiffel Tower.

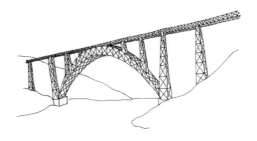

Figure 5-17 Pont Garabit in Auvergne by Eiffel.

Scientific principles and new materials

Jens Larsen

Figure 5-18 The Sydney Harbour Bridge.

the time. Eiffel, on the other hand, fully embraced the new structural forms, clearly expressing that his bridges were supported by steel trusses, rather than hiding the structure behind a façade. His philosophy was clear; the solution that is suitable for a statue, where the *impression* is paramount, is not applicable to a major bridge, where the *function* governs.*

More or less at the same time, in the 1870s, in Great Britain, design was progressing under the leading British bridge designer of the day, Thomas Bouch, on what was to be the largest span structure in the world: the railway bridge over the Firth of Forth near Edinburgh, Scotland. During this design, an appalling tragedy occurred, when the recently completed Tay bridge, some miles to the north, collapsed in high winds, killing 75 train passengers. The Tay Bridge, also a Bouch design, was a cast iron structure of 85 spans, with the rail-line being some 25 m above the river Tay. The collapse occurred primarily because the designers had failed fully to consider the effect of wind loading on a light metal structure. For years, bridge builders had been used to masonry structures,

where the self-weight was so massive that even the highest winds acting laterally on the structure of the bridge could not cause it to overturn. With a much lighter iron bridge, that was not necessarily the case. Unlike Eiffel, whose bridge piers and towers gracefully splayed out towards their bases to provide extra stability against wind, the designers of the Tay Bridge had used narrow vertical trussed columns. It is clear that the design of the Tay Bridge, using the new material and structural form, was not appropriate to the *magnitude* of wind loading.

It is not surprising that in the outcry after the Tay Bridge collapse, Bouch was thrown off the job. The commission was instead given to Benjamin Baker. Baker commissioned the Astronomer Royal to calculate likely wind loads on the structure; his conservative response, entirely understandable at a time of national outrage, was a figure perhaps 100% higher than that which would be used today, but around 400% higher than those used in the nominal design of the Tay Bridge.

Bouch's original design for the Forth Bridge was a relatively light and slender suspension bridge. Such forms were increasingly popular in the USA at this time for long spans, since they

Andy Tyas

Figure 5-19 The Forth Bridge, the two large spans.

*Of course, most *buildings* fall somewhere between these two simple extremes. Balancing appearance and structural function is, as we shall see in more detail later in the book, one of the key challenges facing any design team.

Figure 5-20 The Forth Bridge – there is a clear distinction between the tension and compression members.

the lattice structure of the cantilevers and joining spans were identified and designed differently; compression members are solid and stocky, to resist buckling; tension members are open and light to reduce wind resistance where possible. Compression members were intricately braced together, again to help resist buckling.

However, beyond this response to wind loading, another feature is clear. The overall structural behaviour of the cantilevers is clearly expressed in their form. The cantilevers taper longitudinally from a huge depth at their foundations to a much reduced depth at the tips, exactly following the reduction in bending intensity from the root of a cantilever to its tip.* Reinforcing this idea, the size of the members making up the compression and tension arms of the cantilevers increases towards the supports. Even the joining spans between the tips of the cantilevers are designed to reflect the distribution of bending forces, with a larger depth at the centre than at their support points on the tips of the cantilevers, following the bending moment diagram of a simply supported beam.

The Forth Bridge has rightly become a famous landmark, and a monument to the technology of the time. Often it is quoted as an architectural marvel, yet the entire form and its details seem grimly, almost joylessly, developed to answer the structural challenges of high wind and long span only. It feels as if elegance, subtlety and aesthetic appearance have knowingly been left out of the design concept. It almost feels like a testament to human ability to counteract Nature's challenges. Perhaps it is this that brings the admiration, not the aesthetic qualities and proportions of the developed structural form.

primarily employ highly efficient tensile action, thus minimizing the self-weight of the bridge. This is often a critically important consideration in long span structures, and the Forth Bridge was to be a colossal size, with two main spans of 524 m. The form of Baker's bridge, however, was dominated by his response to the shock of a wind-induced failure; it was a form which seemed to eschew lightness and scream "stability!".

In place of a light suspension bridge, he proposed a cantilever structure: three colossal steel frameworks founded on shallow foundations close to each shore and in shallow water at midstream. From these foundations, the cantilevers spanned outwards in both directions, tied down by counterweights in towers at the shorelines, and reaching to meet each other at their far tips, from where shorter spanning lengths bridged the gaps between the cantilever tips. Each of the three cantilevers is clearly and deliberately tapered outwards towards the base to increase its resistance to overturning. Members experiencing tension or compression forces in

*The same practice that Nature adopts in tree trunks and branches – also cantilevers.

Scientific principles and new materials

Half a continent away, at the same time, the Habsburg Empire was nearing its peak. Hungary had recently been elevated to joint over-lordship with Austria over this huge dominion. Budapest, the second city of the empire after Vienna, was experiencing a confidence, stability and wealth which it had rarely known. 1890 was to be the thousandth anniversary of the founding of Hungary by St Stephen, and the city had a new spring in its step. The first Metro system in continental Europe was under construction, a new Opera house had recently opened. As a centrepiece a new bridge, the Szabadsag Bridge (pronounced "Shobadsag") was commissioned to provide further links between the two halves of Buda and Pest, separated by the Danube.

Figure 5-21 The Szabadsag Bridge in Budapest.

Figure 5-22 The Szabadsag Bridge in Budapest, close-up of the widening and thickening of the top flange of the cantilevered beams.

The new bridge was not to be on the same scale as the Forth Bridge, having a main span of around 220 m, and would not be as heavily loaded, although it was designed to carry a new tram-line. The idea of using cantilever structures for long-span, heavy bridges was now becoming accepted, and a form similar in principle to the Forth Bridge was chosen: a pair of lattice cantilevers stretching from close to each bank, joined at the centre by a structurally independent span. But the shorter span and somewhat lower loads and the lack of the immediate shock and concern over wind loads that was so fresh in British minds, enabled the designers to be less dominated by purely structural considerations. In particular, a conscious decision was taken to mirror, as far as possible, the slender curve of the famous chain-link suspension bridge, also in Budapest, linking the Parliament and Palace, where the structural support tapers almost to nothing at the very centre of the span. This is appropriate, as long as one follows the cantilevers to their tips; the ideal form of a cantilever is to taper to the tip. However, the joining span is very different. As expressed at the Forth Bridge, structurally the ideal shape of the joining span is a curve becoming deepest at the *centre* where the bending forces are highest. The designers in Budapest were fully aware that they were not following the ideal line of force, and had to make provision for this. Instead of the thickness of the top and bottom members of the lattice-work being able to be reduced at the centre, as in the Forth Bridge, they had to considerably increase their sizes right at the centre to compensate for the extreme lack of structural depth there. Figure 5-22 clearly shows the extra-thick steel plates that were riveted together to provide the extra bending resistance.

Andy Tyas

This is an example where the structural principles were one of the (important) considerations in the design, but not dominating the design. While the Forth Bridge is justifiably famous for its colossal presence, it is arguable that the Szabadsag Bridge is a more elegant and beautiful creation.

The two design approaches described above highlight two different methods. The development of new materials and analytical structural knowledge, which led to the division between engineers and architects, has produced a spectrum of possible starting points for design.

At one extreme is a structurally led design (as in the case of the Forth bridge), where the structural form is the driving force, with all other aspects of the building design being treated as subservient. The other extreme would be to entirely ignore the structural issues until very late in the design process, and have a solution where the structure is dealt with as an add-on, with its role being simply to make the building stand up. Fortunately, few designers are so dogmatic as to take up one of these extreme positions, which makes most building designs and design teams fit in somewhere on the infinitely graded scale between these positions. The art of building *is* and *must be* about a partnership between the architecture and the engineering, and frequently about a *reconciliation* between the conflicting aims of each side. The art of *designing* buildings, in its fullest sense, has to answer the challenges and employ the skills of both sides, in the same way that the pre-Industrial Revolution designers combined that skill in one person.

But if the architect and engineer can develop a mutual understanding, trust and respect, the chances of producing great buildings are enhanced. By working together at conceptual design stage, the two sides can help to produce concepts which develop into final structures that satisfy the original architectural concept and the aesthetic vision, are technically viable and meet the financial limitations within which everyone has to work.

Detailed discussion of recent projects demonstrating this understanding and collaboration is given in the second part of this book. The case studies explain how the collaboration between the architect and engineer can be beneficial to the project. However, for now, we will finish with two simple, yet instructive recent examples, which highlight what the interaction of modern materials, analysis and construction processes can produce in the hands of sympathetic design teams.

The first is the new building at the University of Utrecht in the Netherlands, by celebrated Dutch architect Rem Koolhaus working with engineer Rob Nijsse (whose work will be investigated in more detail in the second half of the book). The commission was for a multi-function building; a restaurant for 1000 people, two big lecture theatres and examination halls, all within one building.

Koolhaus's concept was for a large two-storey building folded over on itself (see Figure 5-23). The lower floor would contain the restaurant areas, and the upper floor the lecture rooms. The floor of the lecture theatres needed to be sloping and the architect decided to express this form by exposing the floor on the building elevation and folding it over to become the roof in one elegant continuous sweep of a concrete slab. To ensure that this expression was clearly visible, the entire end of the structure was enclosed in a full-height glass façade.

The architect felt that it was important for the appearance of the building to have the curve forming the roof and floor with a constant thick-

Figure 5-23 The new building at the University of Utrecht, front elevation.

ness of 400 mm. Structurally, a 400 mm thick slab was more than adequate for the first floor, which was supported at regular intervals by columns extending through the restaurant area. However, the roof slab was a different matter. It was not possible to place columns *inside* the lecture theatres, so the slab had to span up to 20 m from one side of the theatre to the other. In addition, the roof had to have insulation included, and Koolhaus insisted that this be incorporated in the 400 mm depth so that his vision of a constant thickness curving around the façade could be realised. This reduced the depth available for the load-bearing structural slab to just 200 mm.

It can be easily demonstrated by a relatively simple structural analysis, that to make the slab span 20 m across the lecture room, it would need to have a minimum thickness of 500–600 mm. A 200 mm thick reinforced concrete slab spanning 20 m can be made sufficiently *strong* by using high-strength concrete or large

amounts of steel reinforcement. However, in a slab of this depth, no matter how strong, the *stiffness* will not be adequate. Regardless of the amount of reinforcement used, when loaded the slab will deflect downwards, perhaps by several hundred millimetres, which will cause the underside of the slab to crack.

In a single-span slab, just as in a beam, the bending moments are greatest in the centre of the span, which is where the slab should be strongest. Consequently, the engineers tried keeping the façade line thin and making the slab thicker in the middle, so that the extra depth reduces the bending stresses. But, of course, this increases the weight of the slab in the centre of the span, just where it is least wanted, seriously increasing the bending moments at the slab centre. Just as in the case of the Sydney Opera House (described in the previous chapter), this attempt to put more material in to counteract bending led to an increase in weight and a consequent increase in bending. In this case the approach could have worked, but it would have required a significantly increased depth of slab in the centre to balance the problem, maybe 500–600 mm.

The engineer explained this to the architect, who asked the kind of simple question that perhaps only a non-expert in structural matters would formulate: *"Why do you need this thickness in the centre of the span?"*. Of course, the reason was to resist the bending moment in the centre of the span. The bending resistance of any member is generated by the opposite actions of compressive and tensile stresses acting within the member at a distance apart. Thus, a member with a bigger depth will have a bigger separation distance, and hence will resist bigger bending moments. As a result, a very thin slab is not able to resist large bending moments generated in long span members. At the centre of the span, where the slab is tending to sag, the compression is at the top of the section, and the bottom of the section is in tension. In reinforced concrete members, the compression resistance is provided by the compressive stresses in the concrete, and the steel tensile reinforcement, which is at the bottom face of the concrete, resists the tension. Under high bending stresses, the concrete, which cannot take high tension, will crack.

Again, the architect asks a very non-technical question: *"The lower concrete cracks? Why do we bother putting in concrete if it is going to crack? Why not take the concrete out? There is a lot of weight associated with it, and it is not doing anything helpful."*

This simple question inspired an interesting solution. Nijsse, the engineer, produced a concrete slab with a constant 200 mm depth. Close to the supports, where the bending stresses are relatively low, the bending resistance can be provided by steel reinforcing bars contained inside the depth of the slab. At a calculable distance from the supports, it becomes impossible to provide this bending resistance within the depth of the slab. The normal solution would be to thicken the slab, but Nijsse's and Koolhaus's decision was to keep the slab thickness constant and drop the steel bars out below the slab, strutting them off against the concrete. This allowed them to achieve the necessary structural depth, without increasing the volume and mass of concrete. As Nijsse describes it:

> "The slab is much thicker than it really needs to be at the support; the bending stresses are very low at this edge. We made the calculations to find out where the concrete would begin to crack if it was a normal slab, and from there we suspended the reinforcement below the concrete. And that is very beautiful, where the rein-

Figure 5-24 The concrete slab with a "belly" of reinforcement hanging out in the middle.

forcement comes out of the slab, really a magical moment. It is expressing how the structure works. Rem Koolhaus likes that very much. No fairy tales. Only reality. That is why he disliked the cracked concrete. Why should we pay the building costs for cracked concrete? Take it out!"

The result is a wonderful amalgam of architectural requirement and structural necessity, the mix producing an elegant and structurally honest solution.

The second example is the beautiful and elegant new ski-jump tower and ramp designed by London-based architect Zaha Hadid and Austrian engineers Aste Konstruktion constructed on the Bergisel hill outside the Austrian city of Innsbruck. Ski-jump towers are typically around 50 m high, with the ski-ramp having a slope of around 35° to the horizontal. They are thus quite large structures. A major

Scientific principles and new materials

Andy Tyas

Figure 5-25 The Bergisel hill ski-jump tower and ramp, by architect Zaha Hadid and engineers Aste Konstruktion. Notice the large steel truss supporting the ramp.

practical factor in the design of ski-jump towers is strict constraint on the slope of the ramp. The structure must therefore be very stiff to prevent excessive deflection under the weight of ice and self-weight. This has usually resulted in structures which have closely spaced columns or a structural framework under the ramp itself. Often the columns or frame are boxed in with timber in an effort to produce an appearance which is sympathetic to the forested surroundings. However, the result is typically an inelegant massive monolithic form.

The preliminary design for the new Bergisel ski-jump was for a ramp supported on a concrete beam, which itself was to be supported by three intermediate concrete piers between the base and the main rear tower. As the design progressed, however, a different form evolved. In attempting to lessen the impact of the concrete beam, the designers moved towards a light steel truss. While there was a danger that this would deflect and vibrate under load, this could be counteracted by pretensioning the main longitudinal members of the truss to give it additional stiffness, in much the same way that a taut guitar string is laterally stiffer than one which is slack. This led to a daring step forward. If a prestressed truss could span between intermediate piers, could it not be made stiff and strong enough to span the 68 m from the top of the main tower down to the foot of the ramp without intermediate supports? This would not only produce a more elegant appearance, but also significantly reduce the overall construction time by eliminating the piers. This was a major consideration given that the old ski-jump had to be demolished and the new one constructed in a single spring-to-autumn period.

The engineer's calculations showed that a single-span truss could indeed be made stiff enough, and this was the form chosen. The two main supporting trusses for the ramp deck run longitudinally along the sides of the ramp, and are hidden by a façade. Along the length of these trusses, supports are provided every 5 m or so by inclined struts, which take the load downwards and inwards to a single line of pretensioned cables running along the centre-line of the deck and underneath it. The prestressed cables and the upper trusses thus act compositely to produce a V-shaped truss. The depth of the cables below the deck increases up to the

Figure 5-26 The Bergisel hill ski-jump ramp, viewed from underneath.

mid-point of the ramp, giving a very distinctive "fish-belly" appearance. Structurally, it is so effective that the difference in vertical deflection at mid-span when the ramp is unloaded compared with when it is fully loaded with over 200 tonnes of ice, is just 8 cm.

The resulting form is a breathtakingly elegant and light structural form, reflecting the freshness of its Alpine setting. The designers are rightly proud of their product, calling it "...*an excellent combination of architectural shape and constructional design*," and more poetically, a "...*Toccata and Fugue in F major for the engineer and orchestra!*".

Notes

1   Galileo Galilei, (1638) "Dialogues And Mathematical Demonstrations Concerning Two New Sciences Pertaining To Mechanics And Local Motions".
2   John Ruskin, (1906) *The Seven Lamps of Architecture*, George Allan, London, pp. 70–74.

Chapter 6

Learning from Physical Models: Design Through Experimentation

The previous chapters talked about trial and error on a relatively small scale (in the primitive structures detailed in Chapter 3) and on a much larger scale (in Chapter 4 on Master Builders). However, one can explore through trial and error on a considerably smaller scale using small-scale physical models. There is nothing that will reveal better the structural behaviour of a cantilever or a truss to a layperson than a simple small-scale model made out of simple materials such as cardboard or balsa wood loaded to destruction. Often it is this way of learning that excites young students about structures. The authors use this approach in their teaching with architecture and engineering students. Small-scale models are built and load tested. It is amazing how much the students learn through these exercises. What is more interesting, though, is the enjoyment they get from such a hands-on, practical approach to learning about structural behaviour.*

The exercises with students include simple tasks such as designing and building a small-scale model of a paper bridge, learning about animal skeletons by understanding how they work, designing and building in balsa wood a large-scale model of a clear span structure for a sports hall, or exploring material properties of conventional (concrete, steel, timber, etc.) and less conventional (glass, plastics, plywood, etc.) building materials and designing an object that would celebrate these properties. In addition, students explore through physical models the structural form and behaviour of some advanced systems, which otherwise would be difficult to explain, such as tensegrities, reciprocal frames, membranes, etc.

These explorations, apart from the obvious aim of exploring and learning about structural form, all have an additional emphasis. Some are about trying to get a fast intuitive reaction (as for the paper bridge where the students have no previous warning, and the finished model needs to be built in 40 minutes), while others are about team working (for the sports hall mixed teams of architecture and engineering undergraduates are expected to come up with a design that not only carries the design load, but is an equally elegant and imaginative structural form).

It is not the aim of this book to go into great detail in presenting these student explorations using physical models. One can come up with an endless list of useful exercises that could aid

Peter Lathey

Figure 6-1 Building a paper bridge model in 40 minutes.

*It is always after projects like this that architecture students ask for more "structural explorations".

Learning from physical models

Peter Lathey

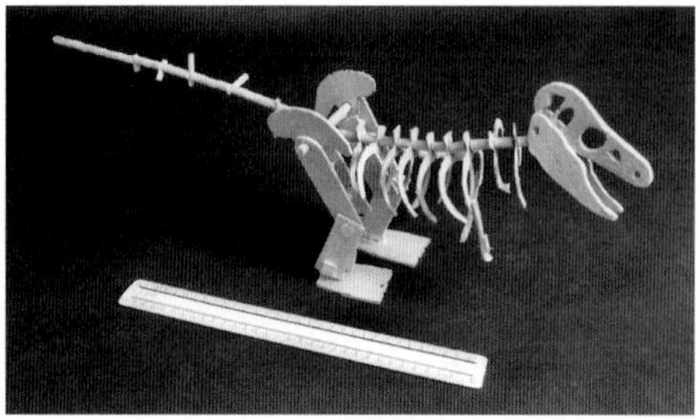

Figure 6-2 The "animal magic" workshop: producing structural skeleton models.

the understanding of structural behaviour and form. However, it is important to state how much these help in understanding how structures behave. The (often) very abstract equations and numbers become meaningful and bending, compression, or tension members become easy to imagine. In a hands-on, almost playful, way, things that at times seem only comprehensible to the mathematically gifted ones, become obvious to everyone. The challenge of designing a structure translates into sketching, exploring

with physical models and trying new things to make these small model assemblies of structural members stand up.

Architecture students love this method,* because it helps them more easily understand very technical issues, ones that are often beyond their technical abilities. Equally, physical modelling is beneficial to engineering students who through their education get a good grounding in analytical methods, as well as the use of sophisticated structural analysis software. It is worth stating here that the authors are not trying to diminish the importance for structural engineers of having mathematical and analytical skills, and being able to use and develop complex CAD and computational structural analysis software. Without these skills, modern day structural engineers would not be able to do their job. We are only stating that one can "learn" a lot about structural form through small-scale physical models. The lessons can be invaluable even for the most mathematically gifted because the models can convey issues about structural form in an obvious and easy-to-understand way.

Peter Lathey

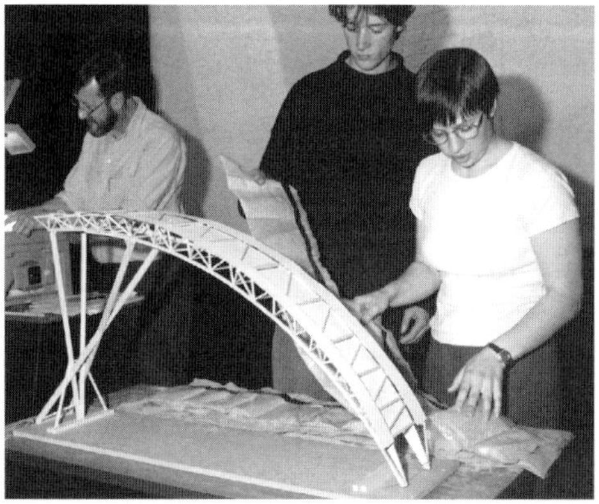

Figure 6-3 Joint architecture/engineering project: clear span structures.

Figure 6-4 Learning about reciprocal frames through physical models.

Peter Lathey

*It is not surprising, as many of the students come into university without *any* mathematical background!

That in effect is the aim of this chapter: to show how physical models have been used to understand structural behaviour and to create imaginative structural forms. It is impossible to say when physical models were used for the first time as an aid for understanding structural behaviour. Models have been used as a means of representation of what a design might be like* and, also, as in present times, in developing architectural ideas.

There is evidence however, that models have been used in exploration of structural behaviour and in developing new structural systems. Wallis, the Cambridge academic and scholar, writes[1] how in the period 1652–53, he built models that helped him understand the structural behaviour of singular and multiple (reciprocal)[†] grids (the singular are very similar to the model presented in Figure 6-4). These structures consist of mutually supporting beams which at the inner end support each other and at the outer end are supported by an external wall or column. The structure forms a polygon in plan with the number of sides equal to the number of beams (as presented in more detail in Chapter 4). Only by building small-scale models of reciprocal frame structures, could Wallis understand their structural behaviour. In addition, his investigations looked at the geometry of the system and defined the parameters of the structure. These explorations would not have been possible without the use of physical models.

Anthony Gaudi

Wallis was not the only one who used models to explore structural form. One can justifiably say that Anthony Gaudi, the Spanish architect, was one of the pioneers in using physical modelling to devel-op his imaginative, and also extremely complex, forms. On some of his projects, he worked closely in a team with a sculptor and an engineer. In developing his architectural solutions, Gaudi, was guided by the configuration of forces and by the structure of organic forms. Gaudi's buildings are often concrete manifestations of diagrams of forces. He used graphic statics and funicular models to develop his imaginative forms.

Funicular models are models that use suspended catenary networks with hung loads at various points. If inverted and "frozen" these models give the most efficient structural form for that particular loading case. Translated into a simple form this means that if we take a piece of string and hang weights along its length, the string will deform. If we are somehow able to make the deformed shape of the string permanent – "freeze it" – and turn it upside down, the arch-like structure will have the most efficient structural form for that loading case (the previously applied weights). This method can be used for finding much more complex linear forms, as Gaudi did. Frei Otto and Heinz Isler, as explained further in this chapter, also used funicular models for creating efficient structural forms.

Gaudi arrived at the final structural shape of Santa Coloma church by using this method. It is interesting to note that the physical model of the inverted catenary arches form of Santa Coloma was used for construction by direct measuring from the model. This is probably the first time structural models were used, by scaling up, to construct a building. In addition, the structural form derived through physical models influenced to a great degree the final appearance and defined the architecture of the building.

Figure 6-5 Basic funicular model.

Figure 6-6 More complex funicular model showing the development of a cathedral section.

Figure 6-7 The final structural form of Santa Coloma.

---

*Byzantine fresco paintings, for example, often present the patron of the church holding in his hands a model of the church.

[†]At the time he only refers to them as grids. The term "reciprocal" is contemporary, see Chapter 3 on primitive structures.

Learning from physical models

"What I had seen in Barcelona was a work of a man of extraordinary force, faith and technical capacity ... Gaudi is 'the' constructor of the 1900s, the builder in stone, iron and bricks ..."[2]

In the above quote, Le Corbusier describes Gaudi as an architect of great practical and technical ability. One can like or dislike Gaudi's expressiveness of form, but it would be difficult to disagree that it is extremely rich. It is interesting that Gaudi developed a real form of team working in developing his schemes. He worked closely with craftsmen, sculptors and engineers, who assisted in translating his sketches into physical models. He felt that it was important to have a real feel of what the material's properties are and how one can work with the material and use its properties to create the imaginative and expressive forms.

Anthony Gaudi came from a family of a coppersmith and a boiler-maker. One of his uncles was a wood turner. Coming from a family of manual workers helped Gaudi develop a feeling for working with materials and using crafting tools. That is probably where his first interest in making things (and architecture) came from. He studied architecture at the newly, at the time, established University in Barcelona. He was there

Figure 6-8 Park Guell, Barcelona, entrance pavilion.

Figure 6-9 Park Guell, the colonnade.

from 1873 until 1878, which, in many ways for him was not an easy time. During this time his mother died and he had to interrupt his studies to fulfil military obligations. In addition, his father was not earning well and Gaudi was under constant pressure from home to work so that he could help the family and support himself. Therefore, it is not surprising that he did not always do well at university and that sometimes he had to resit exams. However, small architectural jobs (to start with) helped him not only financially but also with valuable experience and contacts.[3]

Gaudi created a great number of very interesting buildings. The private residences, churches, Park Guell, the unfinished Sagrada Familia are all examples of great imagination and of an extreme richness of form. His inspiration came from the world of Nature, the Catalan arts and deep Christian faith. Complemented by his ability to work with materials in a sculptural way and his interest in geometry and transfer of forces, this made a mark on his architecture.

Gaudi experimented with forms. His experimentation was often beyond the analytical skills of the engineers he worked with. For the design of the Church Santa Coloma he worked closely with the engineer Eduardo Goetz and the sculptor Bertran. Gaudi spent a long time developing a catenary model for the church. This involved designing a wire model with small

Figure 6-10 Park Guell, the colonnade.

Figure. 6-11 The unfinished Sagrada Familia church in Barcelona.

weights hanging at various points. By inverting the model and replacing the resulting tension forces with compression forces, he could get the most efficient structural form. The physical model was extremely complex, but one that allowed innovative structural forms to be created. It allowed the architect, engineer and sculptor to work together in a team and to create structural forms that could not previously have been imagined or created because of their complexity.

It is interesting how Gaudi came up with this method out of necessity. The method was new* in the sense that it had not been applied before, and Gaudi was the first one to use it in developing conceptual structural design. Although Gaudi worked with an engineer, the complexity of the forms would have been impossible to deal with in an analytical way and his imaginative forms would not have become reality without the use of models.

Apart from experimenting with forms and learning about structural behaviour, Gaudi's small-scale models helped him come up with the most complex, yet efficient, structural forms. His beautiful tiled hyperbolic roofs, like the one on the Sagrada Familia Infant School, the parabolic arches on the top floor of Casa Mila, and the pointed arches in Park Guell, are all examples of efficient structural forms translated into architecture.

Most architects would agree that Gaudi challenged the existing architectural forms. All his designs are very powerful, expressive and very different to any architectural style. He was able to take structural morphology (a more sophisticated term for structural form) and architectural form beyond known limits by being able to

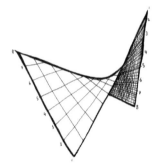

Figure 6-12 Hyperbolic paraboloid shape.

work with the material properties and in a team of people having similar ideas, but looking at the problem from a different point of view.

Buckminster Fuller

Buckminster Fuller was the first to use *geodesic geometry* to create extremely efficient structural forms on a large scale – geodesic domes, he was the inventor of the *"tensegrity"* idea and one of the pioneers in the field of *deployable structures*. It is extraordinary how wide his scope of research was: cars, sustainability issues, geometry and prefabricated systems are only a few of the issues he explored.

"Architect, mathematician, engineer, inventor, visionary humanist, educator, inspirational orator, and bestselling author, R. Buckminster Fuller has been rightly called the 20th century Leonardo da Vinci."[4]

This is how Baldwin describes Buckminster Fuller in the preface of his book about Fuller's work. The emphasis of this section, however, is to show how he used physical models in his structural explorations.

Buckminster Fuller was born in 1895 in Milton, Massachusetts in a family where most members were educated. His father was a merchant, but in the family there were a number of lawyers, editors, public speakers and other distinguished members of society who traditionally studied at Harvard.

From early childhood, Buckminster Fuller was interested in how things worked and he loved making models. His unorthodox education (he was expelled twice from Harvard) consisted of working in a cotton mill in Canada, being part of the Navy during World War I, and doing night classes in market distribution, refrigeration and

---

*In 1748 the method had been mentioned in the writing of Giovanni Poleni, an Italian theorist and mathematician, but it had not been applied. Also, it is unlikely that Gaudi knew about it.

accounting, as well as working in a business with his father-in-law who was an architect. This extremely rich life experience gave him a great breadth of skills and knowledge that no university would have been able to teach him through conventional education.

From very early days Buckminster Fuller was interested in the future of human society. He learned from Nature where nothing was wasted. His goal was to achieve efficiency that would allow for saving of resources so that people could live better. He was the first to talk about and investigate sustainability. Fuller's explorations range from the very practical/technical and problem solving – such as the dymaxion geometry, the dymaxion house,* his design for cars and his quick-built prefabricated houses, geodesic domes and tensegrity systems – to issues that are attempting to resolve problems such as energy consumption, pollution and how to "roof" whole cities.[†]

In terms of structural explorations, Buckminster Fuller's greatest contribution is the design of geodesic domes and tensegrities. Always concerned with both material and structural efficiency, he used domes in his early prefab housing schemes because of all possible shapes spheres contain the most volume with the least surface. A dome has a circular footprint, and it is well known that the circle encloses the greatest area with the least perimeter.[‡] The minimal surface of the dome presents the smallest area through which to gain or lose heat. This makes it a very energy efficient form.

During his research into cartography and development of the dymaxion projection map,[§] Buckminster Fuller realized that the "geodesic line" is the shortest distance between two points on the globe, which has been used by navigators. Also, geodesic forms are evident in Nature. As Nature always does things in the most economical way, Buckminster Fuller knew that the network of geodesic lines should provide the geometry for the strongest and most material-efficient structural system possible.

Through small-scale models, initially, and later models built by his numerous students, Buckminster Fuller's ideas have led to the design and construction of over 200,000 geodesic domes. They range from small-scale buildings, prefabricated systems for housing, to long span and deployable structures. The American pavilion for the 1968 Montreal Expo was among the first long-span geodesic domes with a clear span of 76.2 m. It was a truly amazing structure, with a monorail passing through it, that in the real sense of the word celebrated the development of new materials and structural forms. Unfortunately a couple of years after the opening, the structure was damaged in a fire. It is not surprising that its refurbishment and reopening in 1995 was warmly welcomed by the citizens of Montreal.

Most historic structures (as presented in Chapter 4) worked almost without exception in compression or bending. This is because stone and timber were the most commonly used structural materials. Single stressed members in struc-

Figure 6-13 Small-scale geodesic structure.

---

*The word "dymaxion" comes from dynamic, maximum and iron. It was meant to signify houses that are mass produced, and are also very efficient and low maintenance. Although the dymaxion houses were to be made of aluminium, rather than iron, the title remained unchanged.

†Buckminster Fuller had ideas of how he could build a geodesic dome over New York. He claimed that it was economically viable and structurally possible.

‡If you compare the area that a square with side 1 cm long and a circle with diameter of 1 cm would enclose, the latter is more than three times greater.

§The dymaxion projection map is a flat presentation of the Earth with minimal distortions, using geodesic lines.

Learning from physical models

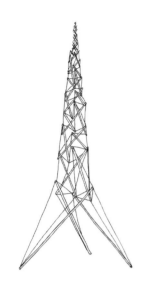

Figure 6-14 The first tensegrity sculpture by Snelson.

tures (in pure tension or pure compression) are always more efficient than structures that work in bending (which is a combination of tension and compression). This is why structures such as trusses with members that are subjected to pure tension and pure compression are very efficient and can span long distances with the use of relatively slender structural elements (as explained in detail in the previous chapter).

The historic compression structures (as explained in Chapter 4) were both beautiful and structurally clever forms. However, the use of tension structures was only possible with the development of new materials (see Chapter 5). Yet, tension structures have an inherently great load resistance and are more efficient than compression structures. On the other hand, compression structures are limited in being able to take very small tension stresses.* Therefore, if one is concerned only with structural efficiency[†] it would be best to design structural forms that use members that are subjected to pure compression or pure tension, and the more of the latter the greater the efficiency.

With this in mind, Buckminster Fuller, together with his students, started experimenting by building various physical models that applied these principles and utilised tension and compression members in structural assemblies. These structures are different to trusses, which commonly would have rigid members, some working in tension and some in compression connected together to form the structure of the truss. In Buckminster Fuller's experiments, the tension members were cables that held up the

compression members and gave integrity to the whole system. These new forms were tensegrities.

Probably the most appropriate definition of tensegrities is *"islands of compression inside a sea of tension"*, given by R. Buckminster Fuller who invented the word *"tensegrity"*, by contracting the words *"tension"* and *"integrity"* – to signify achieving integrity through tension.

A lot has been written about who was the true inventor of the structure because several people simultaneously investigated the possibilities of utilizing such a system. And although the invention of the system itself cannot be attributed solely to Fuller, he is among the few people who came up with the idea of using a new type of structure where the structural integrity is achieved through the tension members. Based on research work done by the Russian constructivists,[‡] one can say that several people, Johansen, B. Fuller, K. Snelson, and D. G. Emmerich, were the inventors of the tensegrity system.

Peter Lathey

Figure 6-15 A model of a tensegrity structure.

---

*It is well known that stone, for example, would fail very easily if subjected to even small bending stresses.

†This of course is not the appropriate approach to design because structural efficiency is *only one* of the many considerations when designing structural forms.

‡Reported in a book by Laszlo Moholy Nagy, *Von Materiel zu Architectur*, first pub. 1968. There, L. M. Nagy included two photographs of an exhibition held in Moscow in 1921, showing an equilibrium structure by a certain Johansen.

It is interesting to mention the role of physical modelling in creating the tensegrity structure: the theory that was developed for the system was based on a background of physical representations, which were the necessary tool for the initial exploration of the system's attributes.

Tensegrity systems are very efficient because the structural members *work* in pure compression and pure tension. By clearly defining compressive and tensile members, the system capitalizes on the efficiency of tension, optimizes the use of compressive members, and eliminates bending altogether. Moreover, the mechanical balance of the system is not dependent on the strength of individual members but on the way the entire structure distributes and balances mechanical stresses.

However, tensegrities are also an inherently complicated system. The system is three-dimensional and therefore the design of the structural form and prediction of the structural behaviour of structure as a whole becomes quite a task. Also, each member within the system is pretensioned, thus there is the threat of progressive collapse in the case of losing the pretensioning. Because of this, tensegrities have not been used in their pure form. However, the tensegrity principles have been applied in the design of long-span lightweight dome structures. The efficiency of using pure compression and especially pure tension structures is the dream of every structural engineer. The lightweight contemporary tensegrity domes, (known as Geiger domes) use tensegrity principles to a great degree today.

It is also interesting to mention that Buckminster Fuller experimented with structures "of Eden" using geodesic domes and a primitive form of ETFE-like cladding. At the time it all seemed like a fantasy. Today his ideas have been developed and used in projects that are truly amazing such as the Eden Project discussed in the second part of this book. In many ways Buckminster Fuller was truly ahead of his time. The work of Buckminster Fuller has had a great impact on many fields. It is difficult to say where his impact was the greatest. However, it is evident that lightweight structures would not have been what they are today without his contribution. His ideas still inspire and influence contemporary structural morphology.

Frei Otto

Frei Otto is another person who used physical modelling techniques in developing innovative structural form. He grew up in a family of sculptors. Both his father and grandfather were sculptors. As a teenager he took up gliding and later, during World War II, he joined the Air Force as a pilot. The last two years of the war he had to spend as a prisoner of war at Chartres where he was put in charge of the construction team attempting to repair damaged bridges. During those two years he first started his explorations on lightweight structures. The war years were a time when there was a great shortage of materials and an abundance of labour. The lack of resources lead Frei Otto to explore minimal

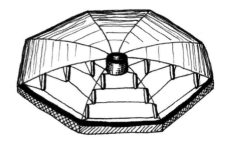

Figure 6-16 The Geiger dome applies the tensegrity principles.

Learning from physical models

structures. During his time in the war camp Frei Otto developed the fish-belly lattice girder.

After the end of the war he studied architecture (1948–1952). This is the first time he realised that his structural ideas were more than pure common sense. Some of them were real innovations. Frei Otto's interest in structures led him to study suspended roof structures and gain a doctorate in the field. Frei Otto is a person who through his education (architecture and scientific/technical doctorate) has bridged the gap between architecture and structural engineering. One can compare him to the great medieval Master Builders who were in charge of both the purely architectural and the structural issues. Frei Otto's contribution is in creating the notion of lightweight structures and developing these structures into aesthetically pleasing architecture. He has truly taken lightweight structures a great step further in their development. It is not surprising therefore that he is well known for his imaginative designs of lightweight structures such as membranes, tree structures, cable nets, pneumatic and kinetic forms.

Frei Otto was very interested in developing beautiful but also minimal and efficient structural forms. Even more than Buckminster Fuller he investigated pure tension and pure compression structures. It is important to note that his structural explorations were almost without exception done by experimenting with physical models.

Like Gaudi, in developing his structural forms Frei Otto used the fact that freely suspended chains or cable nets automatically adopt the most favourable shape. If this shape is stiffened and inverted, it creates the most efficient compression structure for the particular loading case. It is the same idea of catenary

Figure 6-17 Spider web explorations.

forms that Gaudi had used before, but applied to form-finding of membranes and tension structures. The similarity between Gaudi's and Otto's work is that they both use funicular polygon models to develop structural forms which are then translated into architectural form. However, the differences are in the used materials and the actual forms that were developed through these structural explorations.*

Another great source of inspiration for Frei Otto was Nature and the beauty and efficiency of natural forms. He investigated spider's webs, tree forms and soap bubbles. Nature is not wasteful and can teach us both about beauty and structural efficiency. Nature's creation combines efficiency and beauty in the most perfect way.

Frei Otto is a great believer in lightweight and adaptable structures, similar to the ones we find in Nature. In that sense he stated:

"Instead of doing justice to today's needs in colossal arrogance our buildings claim fixed values for an indefinite time. We need buildings that fulfil their task today and will do so tomorrow, which in other words, do not age … and thus become a drag upon the economy as well as the visual environment. But in order to build adaptably we must try to build as lightly, as movably, as possible and with the greatest perfection technically available".[5]

In developing minimal structures Frei Otto used physical modelling techniques. The forms he worked with, such as tension structures, pneumatic structures and tree structures, were extremely complex, and often the only way to deal with the three-dimensional complexity was to build physical models. His explorations with small-scale physical models helped him refine the structural form by investigating and understand-

---

*For his built structures Frei Otto used contemporary structural materials such as steel, ETFE membranes, timber and glass, while Gaudi mainly worked with masonry structures.

Figure 6-18 The Munich Olympic Stadium, by Gunter Benisch and Frei Otto (1972).

ing the structural behaviour. The models were loaded to check extreme loading cases including wind and snow loads. For the purpose, special precise measuring devices were developed to measure the deformations under load and to aid the understanding of structural behaviour.

In the case of membrane structures, once the efficient and aesthetically pleasing form was achieved, the models were scaled up and cutting patterns for the membranes developed. In this way the structures were constructed by direct use of models. It is interesting that the development of mathematical models for analyzing these complex three-dimensional structural forms, in most cases later, confirmed that the forms derived through physical modelling were the most efficient ones. For form-finding of minimal surfaces he used spatial linear configurations of catenaries with high-tension soap films. The properties of such surfaces were important in developing prestressed tensile surface structures.

As early as 1962 Frei Otto built a timber grid shell using physical models for form-finding for Deubau in Essen and in 1971 he designed the famous grid shell for the Federal Garden Festival in Mannheim. The structure was a timber lattice structure with a beautiful and complex form

Figure 6-19 The Munich Olympic Stadium, detail of the membrane.

that spanned 80 m. The form was developed using models of suspended nets.

Frei Otto's great contribution is that he identified processes that create efficient and beautiful lightweight structural forms. He defined methods of creating lightweight structures. His structural explorations into new forms were using scientific and aesthetic principles. With his designs Frei Otto truly bridges the gap between science and art.

Frei Otto's designs have been and continue to be a source of inspiration to many architects and engineers.* His design for the Munich Olympic Stadium roof structure built for the 1972 Olympic games is still among the most impressive stadium designs of the twentieth century.

Heinz Isler

Heinz Isler is a Swiss structural engineer well known for his free form shell structures. He believes in the unity of architecture and structural engineering or as he puts it: *"Architecture and engineering are just two aspects of one thing"*.[6] His view is that architects and engineers should work together in developing structural form. If the cooperation between the two professionals (architect and structural engineer) starts early in the project and is successful, then it is most likely that the design will also be successful. Isler applies these principles to his designs of thin concrete

Figure 6-20 Form-finding of shell structures using small-scale models.

Figure 6-21 Concrete shell structure at service area on A12 motorway in Switzerland designed by Heinz Isler.

shell structures that, as a result, are truly minimal, elegant and sculptural.

Heinz Isler was born in Zolikon, now part of Zurich, and was educated in Switzerland. Although trained as a structural engineer, Isler is a talented watercolour painter and is good at sketching. An exhibition of his paintings representing the Swiss landscape was held while he was still at school. At one point Isler considered starting a career as a painter, but the call to create sculptural concrete shell structures won over painting.

In 1959, at the First Congress of the International Association for Shell Structures, held in Madrid under the direction of Eduardo Toroja, Heinz Isler gave a paper entitled "New Shapes for Shells". In his presentation, Isler presented three ways of creating free-form concrete shells: using moulded earth instead of formwork;[†] using an inflated rubber membrane; and using hanging cloth reversed where the draped fabric defines surface shape in the same way as a hanging cable defines a funicular line. He showed physical models and talked about how they can be scaled up, which gave unlimited possibilities in creating free-form shells. The paper made a great impact because everything Isler proposed seemed so obvious and simple, yet so innovative.

Many of the shells that Isler designed were constructed using these methods. He would build small-scale models using hanging fabric, "freeze" the three-dimensional shape using epoxy resins and then just scale the model up. At the time when Heinz Isler started designing shells this was the only way one could design

---

*To name just one, the Downland grid shell designed by Edward Cullian Architects with Buro Happold Engineers which was shortlisted for the Sterling Prize 2002.

†The same method was proposed by structural engineer Robert Nijsse of ABT for the ground-floor structure, the Dutch dunes level, of the Dutch Expo building for Hanover which is described in detail in the case study section of this book.

them. Computers were not powerful enough to support structural analysis software for spatial structures. Thus, there was no such software developed at the time. Therefore, very precise instruments were used to measure the small-scale models so that the real structures could be drawn in full scale as a scaled-up version of the models.

The shell structures that Heinz Isler designed were not only elegant and sculptural forms, but they were truly minimal and extremely efficient. Designed by using funicular planes they work mainly in compression, thus no stiffening ribs were necessary. Also, because of the appropriate shape and shell action their span to depth ratio is very big. Many of Isler's shells span 30–40 m with a concrete shell only 8–9 cm thick.

It is interesting that today when we have powerful computers they only confirm that the shell forms that Heinz Isler designed by using physical models were correct. This by no means undermines the computer modelling and three-dimensional structural analysis that we are able to use today. It is important for engineering consultancies to have cutting-edge computer software in order to be able to deal with the detailed design of complex three-dimensional structures. However, the physical modelling used for form-finding remains a very powerful tool for exploring structural forms. Even some of the greatest structural engineers like Heinz Isler still use it. Every winter in his gar-den, Isler creates sculptural forms using fabric that he dips in and sprays with water. The illuminated frozen sculptural forms are a beautiful form of architecture, one that is at the same time sculpture and structure. Yet again these ice forms present the endless opportunities in creating imaginative shell structures. In addition they are in fact physical models of structures that can be built using more permanent structural materials.

These, and equally the small-scale models, reveal many of the characteristics of a structural form as well as its behaviour when subjected to loading. Physical models, alongside other design tools, can give invaluable guidance on which structural forms will "work" and how the structure will behave. It is not surprising, therefore, that some of the greatest designers have used them (and still use them) in conceptual structural design.

Notes

1 Wallis, J., (1695) *Opera Matematica*, Georg Olms Verlag Hildesheim, New York.
2 Le Corbusier, in *Gaudi*, 1977, D. Mower, Oresko Books, London.
3 Mower, D., (1977) *Gaudi*, Oresko Books, London.
4 Baldwin, J., (1996) *Bucky Works: Buckminster Fuller's Ideas for Today*, John Wiley and Sons, New York.
5 Drew, P., (1976) *Frei Otto Form and Structure*, Granada Publishing, London.
6 Chilton, J., (2000) *The Engineer's Contribution to Contemporary Architecture: Heinz Isler*, Thomas Telford, London.

The Link

There is a danger that, seen in isolation, the first part of the book, the study of sources of inspiration for structural design, can be taken as a prescriptive list to be studied and repeated. It does not aim to be. The development of design ability in conceptual structural design as in any other field depends on *application* of ideas and experience. The previous chapters have, we hope, pointed the way towards areas in which designers may look for ideas for structural form. We have also discussed how such concepts have been refined and implemented in classic cases. It is vital for the reader to appreciate that these are not simply *historical* lessons, but are equally valid today.

The following four case studies are intended to highlight this. In these recent built examples, we will look at how the design of four contemporary buildings developed from a clean sheet of paper to a final built structure. The most obvious factor running through all of these examples is the close collaboration between architect and structural engineer from the very earliest stages of the design. In all cases, there was a process of iteration from the architect's initial vision for the building to the final buildable form. More subtly, there is clear mutual respect between the two designers. The architects understand, appreciate and embrace the importance of structure to the building. The engineers on the other hand realise that structure is far from everything in the design, respect the architectural concept and apply their technical skills without losing the architectural vision. All the cases under consideration have developed through a dialogue which has led to some form of compromise, but one where neither the vision nor structural form have been compromised. The designers have successfully bridged the gap between the cultures of the two professions to produce uplifting buildings with inspiring structural form that contributes to the whole.

In preparing the forthcoming sections, we have deliberately not emphasised the messages from the first section. It will be more enjoyable for the reader to draw their own conclusions about the designers' search for inspiration. The case studies present the design development and refinement and how the concepts were translated into reality.

All the case studies follow a similar format. All were based on interviews with the designers. It is perhaps instructive that, in every interview, we were caught up in the designers' enthusiasm for their particular project and for the profession in general; in every case, we quickly threw away our prepared list of questions and entered into an enjoyable discussion of the detail and wider philosophy of design. To try to give a flavour of this experience, we have kept the case studies as close as possible to a verbatim transcript of our interviews, with general background information provided to fill in the gaps. We hope that the reader obtains as much enjoyment from reading these as we have from speaking with the designers.

Case Study 1

Clearwater Garden at Chelsea: a striking roof structure used to bring awareness of global water shortages and sustainable development. By Sarah Wigglesworth Architects and Jane Wernick Associates

In recent years we have been faced with environmental issues on a daily basis. Pollution and waste production go hand in hand with the development of consumer societies. These cause unwanted changes in world climate as a result of global warming. There are areas where frequent flooding has caused material damage and disruption and loss of human life. However, the shortage of water is an even more unwanted result of the developed world. There have always been dry regions, but unfortunately it is evident that an unsustainable way of living has resulted in creating more droughts and floods than ever before. Therefore it is extremely important that human societies take environmental issues more seriously and find ways to tackle them in a more proactive way.

This case study is about a design that does exactly that. The clients for the project, the Americans Beth and Charles Miller, dedicated promoters of sustainable development, had decided to raise awareness of environmental issues and especially the problem of global water shortage by bringing their message to a large audience. They chose the Chelsea Flower Show, the showcase of world gardening, as the place to do it and commissioned Sarah Wigglesworth Architects* in collaboration with Mark Walker Landscape Architecture to demonstrate water recycling and conservation practices through a design for a show garden. The practice had experience in working with sustainable materials (for example, they have used straw in a novel way in several projects) and had researched the use of reed beds for some of their previous designs. They approached Jane Wernick Associates[†] to help them with the structural issues; Clifton Nurseries took care of the procurement and various other people were also involved. The pavilion, which formed part of the garden design, was fabricated and erected by ISV Metcalfe Ltd.

In the design for the pavilion, the architects' approach was to produce a visually exciting structure that would attract the audience and convey the message about the potential for water recycling and conservation in suburban areas. The design was conceived as a garden belvedere or folly, the last outpost of the house as it moved into the wider landscape. The

Case 1-1 Professor Sarah Wigglesworth, Principal of Sarah Wigglesworth Architects (SWA), Project Architect on the Clearwater Garden Pavilion.

Case 1-2 Jane Wernick, Director of Jane Wernick Associates, Project Engineer on the Clearwater Garden Pavilion.

---

*Sarah Wigglesworth Architects, with Sarah Wigglesworth being a principal of the practice, is interested in the making of buildings by exploring the use of readily available materials in a highly innovative way. The best known example of this approach is the Straw Bale House and Quilted Office in Islington, which has been featured in many architectural journals.

†Jane Wernick is the Director of Jane Wernick Associates. She worked for Ove Arup & Partners as an Associate Director where she was responsible for the scheme design of the London Eye. In 1998 she founded her own practice. Jane Wernick Associates specialises in collaborating with architects and the development of innovative or unusual structural forms.

Case Study 1: Clearwater Garden Pavilion

imaginary house, which was responsible for creating the waste grey water, was to be positioned behind the straw bale wall. Behind the wall was also the formal garden belonging to the house. The folly and the reed beds where the water recycling took place were imagined to be on the margin of the estate.

This project is different from others because it was intended for the Chelsea Flower Show where the design has to be approved one year in advance of the garden going on site. An additional requirement was that the project had to be constructed over a short period of time and dismantled after the end of the show. All these requirements had an impact on the design approaches adopted by the team. Sarah Wigglesworth explains:

"We were first appointed in early 2000. The project was about two years in its gestation partly because we had to have a design up and running in order to apply to be admitted to Chelsea at least a year in advance. We were trying to go for one of the largest show plots so it was important for us to make the plans carefully."

The initial idea about the pavilion was very different to the final design. Wigglesworth recalls:

"It started in a very organic way. That was when it was really out of control. It started organically because that was part of the landscape strategy. Originally the pavilion was going to be something that would completely blend into the landscape. In the course of

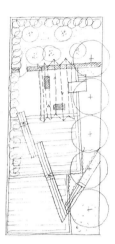

Case 1-4 Architect's sketch of the final garden design: plan.

designing the garden over quite a long period of time the pavilion emerged as something a lot more identifiable as a structure within a landscape rather than a part of the landscape."

Jane Wernick adds:

"In the beginning you were thinking about something like a hut, a willow hut. We discussed that at our first meeting".

Wigglesworth continues:

"After that, I took a different approach and came up with the model for the folded roof. From then on it took a completely different direction. I had an image in my mind, which came from the initial conversation we had, which was to do with layers and glass. Material considerations then came in."

This new concept was the one that was developed and later executed. It was a canopy supported by tree structures on a non-regular grid. The architect's vision was to use trees that deliberately looked different to real trees. They were supposed to look like something man-made or manufactured, and they needed to act as the supporting structure for the grid carrying the folded plane canopy. This is where the real challenge started.

Case 1-3 Initial design idea for the pavilion.

Case 1-5 Model of the final pavilion roof.

When one looks at a growing tree, it is always wider at its base. The tree trunk and branches become smaller in section towards the top of the tree. This is because the tree acts as a cantilevering structure with the largest bending moment at its base. In addition to this, a real tree trunk in essence is a non-regular cylinder, with the branches being smaller cylinders attached to the tree trunk. This form provides the buckling resistance against the vertical loading from the upper part of the tree and also bending resistance against the wind. Nature has put a lot of logic into the structural form of a tree.

The architect's vision of the tree was very different to this. It was imagined to be of flat sheet cut-outs. Sarah Wigglesworth explains:

"This idea [using plates for the trees] was not driven by engineering at all. I had an image of the Magic Roundabout, the cow and these flat trees made of cutouts. The thinking about the trees, did not come from anything to do with the project, but was just a crazy image in my head."

It is clear that the imagined cut-out trees could have not worked structurally. Under even very small loadings the individual branches or the trunk would have buckled, since they had almost no lateral strength or stiffness. Yet, they were expected to take the load from the grid supporting the folded plate canopy. This is where the engineer came in. Perhaps the easiest would have been to change the trees into tubular structures. This would have provided the necessary buckling and bending resistance, but the architect's vision would have been lost for good. This was a key moment for the project and the start of a real collaboration in the team. Jane Wernick explains:

"From the beginning Sarah was trying to describe her aspirations for this, whereas I was trying to absorb and was thinking how can we achieve that. And sometimes Sarah would say 'That was not what I was thinking at all' and then we would discuss that. One instance of this is when she drew the flat plates and I said they would buckle, but that forced me to think about making a section out of plates rather than rolled sections or extrusions. That was quite a key moment. That gave us a chance to have the tapered sections but still have plates. By taking construction principles like the cantilever needing to be thicker at its base it all started to work with the image."

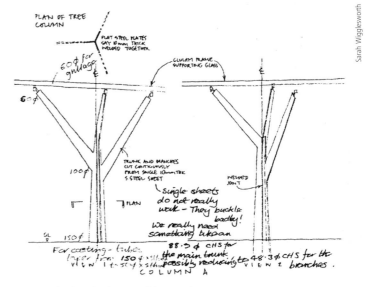

Case 1-6 Design sketches: development of the tree column concept.

Case Study 1: Clearwater Garden Pavilion

Sarah Wigglesworth comments:

"The fact that Jane took that on board and offered solutions to it [the flat planes] was absolutely fantastic."

This is often the root of problems between engineer and architect. The architect has a concept, an idea that he or she wishes to see realized. If the concept is lost, the whole vision of the project disappears. However, the architectural idea may sometimes have little to do with structural "correctness", particularly on smaller buildings where the scale allows some deviation from ideal structural principles. Engineers on the other hand are trained to think in terms of structural efficiency* and are often searching for the most structurally suitable solution to a problem. Put this together with an architect for whom structural issues may be a hindrance to the main aim of the design, and the combination can be explosive. The best architect–engineer collaborations defuse this problem through mutual appreciation and respect for the aims and constraints of the other side.

The whole design process of this project was like that: it was a true team collaboration of equals. The architect was interested in the technical issues and encouraging the engineer to take part in developing the concept. The engineer and the architect worked together exchanging ideas throughout the design process. Thus, the final design became an amalgam of the joint aspirations and ideas. The structural "correctness" was always part of this dialogue, but never became the leading factor in the design. Wigglesworth talks about their collaboration:

"We worked in very much a hands-on way. We sat together doddling away, in a very equal manner. Often when I have worked with engineers it was not like that, partly because of time constraints where they could not give as much time as Jane gave. You work at a distance, you use faxes, and talk on the phone, but there are gaps in the communication when things do not quite come together. I think sitting together is very important."

Wernick adds:

"I do not think we worked that many hours on the project. It was that we worked together very intensely. There were probably three or four really key meetings of an hour each. So we worked quite quickly in a way. You are lucky if you find someone who works at the same pace as yourself. So you can bounce ideas back and forth. I am lucky in that there have been a number of architects with whom I have a good collaboration and who will come and speak to me right from the beginning. There are others who do not. The chemistry has an awful lot to do with it. I am more likely to get involved if I feel welcome in the process. That is very important."

The mutual respect and the designers' ability to go beyond the boundaries of each other's profession enabled them to realize a design true to the architect's imaginative vision with a structure that was also appropriate in engineering terms. Wigglesworth explains:

"We are both quite used to thinking about the other discipline and we both had an interest in collaborative ideas as a theoretical notion. It is about pushing the boundaries. My grasp of structures is basically intuitive. [But] I am not afraid of asking. If you assume that you know what the solutions can be, you do not ask those dumb questions so you do not explore areas that could be legitimately an avenue of exploration."

Wernick comments on the issue of good communication and how to encourage it:

"We need an interest in the other's profession. Early on I was lucky to work with engineers who worked with architects, and I thought that that was fun, and interesting. I think that engineers need encouragement to speak up and ask architects what they are trying to do. The education of engineers needs to change to encourage this.

Case 1-7 Design development: the importance of sketching and doodling.

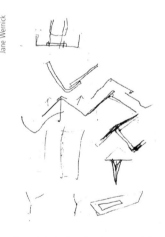

Case 1-8 Design development of the folded plate: the importance of sketching and doodling.

---

*There is an American saying that an engineer is a person who can do for 1 dollar what any fool can do for 2.

"It is important to find the language of communication. I teach architects. Architectural students tend to put their knowledge in different boxes and they have good technical understanding, they have good spatial understanding, and I am sure that the same is true with engineers. I did not learn any architecture at university. I've learnt it all after leaving university. We are surrounded by architecture and there cannot be many engineers who know nothing about it."

The design changed and developed with the involvement of both sides: architect and engineer. There were different driving forces: technical for the engineer whose main aim was to make things "work" without compromising the architect's concept. The architect dealt with the appearance of the structure which needed to be true to the concept and fit the context and use. It is interesting how these different constraints were dealt with and resolved into the final design, for example when deciding on the cladding for the canopy which started off being all glass and ended up as a combination of glass and ply.

For lightweight roof structures, often the biggest load is wind.* A heavy roof, such as a slate roof, will be secured against wind uplift by its dead load (which in this case is the weight of the slates), whereas if a light roof structure is not tied down it could simply fly away when exposed to stronger winds. Wernick explains:

"It was all getting a bit complicated. You have to think about wind uplift and the roof had all these tie downs. Then we had the scheme costed and it was too expensive. I do not think it was an issue how much money they had, but what was a reasonable amount we thought we should spend, because it was only going to be there for a couple of days. We had a big meeting and we decided something had to go. And I said

then that it would be easier if the cladding was partly plywood. And it turned out that it was better."

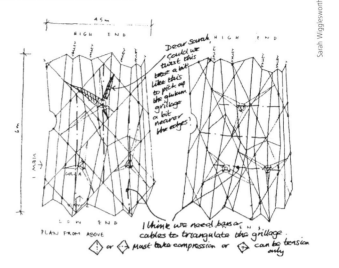

Case 1-9 Communication between designers: engineering notes on architect's sketches.

Case 1-10 Glass roof, plate beams and trees.

---

*That is why for tension structures, for example, where the cladding is just a thin membrane and is very lightweight, it is extremely important that the structure is well secured by tension cables which tie it back to the load-bearing structure.

Case Study 1: Clearwater Garden Pavilion

For completely different reasons, Wigglesworth felt that perhaps the all glass canopy was not the best solution. She recalls:

"I was also having my doubts about it. I was concerned about leaking because it was a series of glass gutters. I didn't like the ties and also, the construction was getting complicated. Finally, it was likely that it would receive a lot of sun and we thought it would be better if we created something like louvres that were directed to protect against sun. It worked out well that way."

Wernick adds:

"It was much simpler to put together without the ties which were a bit of a nightmare. We had them originally because the glass needed to be tied down. But we knew we were not happy with them and wanted a better solution."

The final solution is a lot more elegant without the ties. Also, the ply panels shade the area underneath them and create a pleasant sitting space.

It is interesting that the whole design was developed through models. Physical models were used to start with. The architect made a small simple folded paper model to work out the design and later this became useful to explain to the engineer how the canopy should look. Later, CAD models were developed. Physical and CAD models can be very powerful design tools (see Chapter 6). They can help in the development of an appropriate structural form. In addition, as in this project, they can help the dialogue and communication between the two designers, which in turn aided the collaboration in the design.

Wigglesworth talks about the beginning of the project, and how the use of physical models came about very soon after she approached Wernick.

"I am absolutely useless with structures. I know nothing about them at all. I am very unconfident with dealing with structures and my approach is totally, totally intuitive. I am useless with numbers, I just do not have a clue."

Wernick comments:

"I think an intuitive approach is absolutely fine. We live in a physical world: we know how a table works and you *started making little models*. You wanted something non-regular. Perhaps that's why you were nervous because it was non-regular."

It is important to emphasize how much physical modelling can be instructive in understanding structural behaviour. It is a non-mathematical way of visualizing a structural form. In the case of this project the physical models were used as a tool in developing the design ideas. Thus from the moment the architect built the first model everything was done with models. Wigglesworth recalls:

"The whole thing was done through physical models in the beginning. We could have not conceived it without them."

Case 1-11 Final CAD drawing of the roof, showing tree branches and lattice, plan view.

Jochen Kaelber

Wigglesworth continues:

"From the moment I changed the design from being organic to flat planes, the cardboard model was taken up by Jochen* and he made it into a computer model. Having done that, he could define the pattern pieces and it became consolidated in three dimensions so that it could be treated (understood) by somebody else. Then he made a further model, a cardboard model and we were finally able to resolve the lattice. So, it was an interesting dialogue of these different types of modelling: some computer modelling, some physical modelling, hand drawings, etc. And then in the end we built models of the trees. We made two versions of them to see how they would look. Each model was a different scale. The trees were quite big 1:10. We kept on modelling right until the end."

Wernick adds:

"Having the models and the photographs really helped."

It is this communication that helped the creation of such an interesting structure, one that could do more than just bringing awareness about water recycling and environmental issues. This small project has its real greatness in a showcase of mutual respect, communication and true collaboration within the design team. An architect who readily admits to having intuitive understanding of structures, working with an engineer who is prepared to accept that structural "correctness" need not be the driving force for a building with a message.

Perhaps, the best way to conclude this case study, is by quoting Wigglesworth's words explaining what made and can make communication between architects and engineers possible:

"…Both Jane and I are not shy about doodling and I think that is an imperative – a key part of the communication. It is difficult to work with people who just talk. I find working with engineers best if they can draw beautifully. The sketches are clear and they are lovely to look at. It is the doodling that is most creative and that is crucial in getting the communication to work."

Case 1-12 Final design of canopy and lattice, physical model.

Case 1-13 Tree structure in its final form, physical model.

*Jochen Kaelber was an architectural assistant who at that time was working at Sarah Wigglesworth Architects.

Case Study 1: Clearwater Garden Pavilion

Case 1-14 The completed pavilion, the result of design imagination and excellent communication.

Paul Smoothy

Postscript

After the end of the Chelsea Flower Show, the structure was successfully dismantled and was taken to a warehouse in Devon, where it awaits its imminent reassembly on a new site in Essex.

The design team (architect and engineer) expressed their gratitude to the client who was generous and extremely trusting. They hope that other clients will be equally trusting and generous and allow enough time for the design development as well as bringing the design team together as early as possible in the project.

Case Study 2

Building With Nothing At All: Glass Structures. By Dirk Jan Postel and Robert Nijsse.

In the very earliest and simplest forms of human construction the functional aspects of providing shelter were paramount. As building became an expression of artistic creativity, and as the expectation grew that buildings should delight as well as serve, aesthetic concepts became increasingly important.

Throughout the history of architecture, one of the most frequently used techniques for expressing the elegance of a design has been to emphasize lightness and space over solidity. This can be seen as highlighting the ability of the designer to control and channel the forces which the building structure must support, directing them in the way that they wished, rather than having to provide unnecessarily bulky structure through ignorance. Examples of this include openings for windows leading to the use of columns rather than solid walls; the development of arches to span long distances rather than using closely spaced columns; the wonderful elegance of the Gothic cathedrals, where the space often dominates the structure; modern use of light trusses rather than heavy beams. What many of these moves have in common is the reduction of the amount of structural material, and an emphasis on the space and lightness of the building.

Over the years, this led to a separation between relatively strong but discrete structural framing, and relatively light and flexible non-structural walls to provide the envelope for the building. This approach combines simplicity of construction with flexibility of use; the structural frame can be prefabricated and erected quickly on site, while the cladding can be altered, and openings for doors, windows, etc. can be incorporated without affecting the load-carrying performance of the building. In this type of construction, the primary structural purpose of the "walls" is to resist lateral forces due to wind pressure. To do this, the wall panels are often strengthened by the inclusion of secondary steel, concrete or aluminium framing which supports the panels and carries the load back to the primary structural frame.

In modern times, this cladding envelope of buildings has increasingly been made from glass, both to provide natural light for the inside of the spaces and to give a feeling of lightness. However, using large expanses of glass simply as cladding is relatively wasteful. Glass has very good structural load-carrying characteristics, with a strength-to-weight ratio which compares favourably with more commonly used structural materials such as steel, concrete and timber. One of the reasons why glass has taken a long time to establish* itself as a load-bearing material is its brittle nature. The material collapses in a non-

Case 2-1 Dirk Jan Postel, Architect and Director of Kraaijvanger Urbis Architectural Practice based in Rotterdam, Netherlands, Project Architect for the Temple of Love and glass bridge in Rotterdam.

Case 2-2 Robert Nijsse, Structural Engineer and Director of ABT Engineering Consultancy based in Arnhem, Netherlands, Project Engineer for the Temple of Love and glass bridge in Rotterdam.

---

*Although its use is becoming more common, it is fair to say that there is still for certain applications a need to convince authorities that there is a sufficient safety margin when using glass as the load-bearing structure.

ductile manner, which in simple words means that the structure when subjected to impact loading (for example, when hit by a stone) or excess load will collapse instantly and without giving any warning. This, of course would be unacceptable for a building of any use. There are technical ways of overcoming this issue, such as using toughened or laminated glass, but unfortunately, there are still preconceptions and lack of confidence in the material properties of glass.

However, over the last couple of decades, increasingly the characteristics of glass have been exploited to produce cladding in which the glass acts structurally. Pioneered by designers such as Peter Rice, glazed wall systems were developed where the glass panel, rather than simply being supported by the secondary framing, acts compositely with a light grid of steel or aluminium members as a large bending panel to resist wind loads.

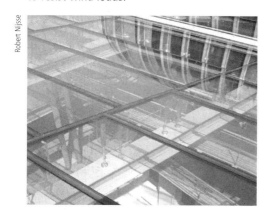

Case 2-3 Andrassy building in Budapest, structural glass floor using glass beams.

Recently, a few designers have taken an even more radical step in this direction. Their reason is as follows: for relatively small structures, with only light loading, why have any primary or secondary structural framing at all? Why not simply let glass walls provide *all* the structural support for both lateral wind loading and vertical gravity loads? Among the standard bearers of this new form of architecture are the Dutch designers, architect Dirk Jan Postel[*] and engineer Robert Nijsse.[†]

Nijsse had had experience of working in structural glass since the mid-1980s, most notably acting as structural engineer for the daring glass structures of Erick van Egeraat's ING office building on Andrassy Street in Budapest in the early 1990s. It was the success of this project, and the fact that a friend of Postel's also worked on the scheme, which brought the two designers together. They found a common interest in the stark lightness of structural glass, which has developed and grown since. Today, Nijsse, Postel and other friends and colleagues undertake regular visits around Europe to find other examples of structural glass and to give talks on their own experiences.[‡]

Uppermost among their experiences has been the difficulty of forging a new form of architecture, convincing clients, suppliers, contractors and checking authorities that their schemes are technically, financially and practically viable. Nijsse recalls the opportunity that the power vacuum following the fall of the communist regime in Hungary afforded him and van Egeraat on the Andrassy building:

---

[*]Dirk Jan Postel is an architect and Director of Kraaijvanger Urbis Architectural Practice based in Rotterdam, Netherlands.

[†]Robert Nijsse is a structural engineer and Director of ABT Engineering Consultancy based in Arnhem, Netherlands.

[‡]For several years Dirk Jan Postel and Robert Nijsse have been teaching at the Masters course on Structural Glass at the Technical University of Delft Department of Civil Engineering.

"We worked on the Andrassy building right after the Iron Curtain came down. There was chaos, no authorities checking anything, so we could use this new idea, of beams made from glass to hold up the roof, and we were able to do something quite new and very special. And this was how we made progress in this area. We showed that it could be done. And now of course you see everywhere around the world they make glass beams, because they have been built before."

Convincing clients to be experimental when money is at stake can be a more difficult matter. Fortunately, Postel and Nijsse found a more than willing client for their first glass structure: Postel's own architectural practice. Their studio in Rotterdam consists of two separate two-storey buildings, which were once part of the city's waterworks before being renovated and converted into offices. A link bridge was required to span the 4 m between the buildings, and Postel saw the opportunity to create a beautiful piece of architecture by pushing the boundaries of technology and using glass to provide the enclosure and structure. Nothing could have been a better advert for the opportunities that structural glass can offer.

The enclosed bridge* that they designed had the deck, supporting beams, side walls and roof made entirely from glass. The only non-glass parts are small stainless steel connections and fittings, which tie the different sections together and provide the mountings for the longitudinal glass beams. The primary load-carrying members are the toughened glass deck and supporting beams, the latter being fabricated from three parallel profiled sheets of glass. The sidewalls and roof, also made out of glass, are present to give enclosure and protection from the environment.

The whole effect is quite remarkable. Seeing from outside someone walk across the bridge is a fascinating and at first even slightly unsettling event, as the walker seems to float on nothing. Looking down from the bridge deck oneself reinforces this feeling of nothingness; there is none of the solid, opaque structure that we expect when crossing a span. Yet the experience feels entirely secure, perhaps due to the enclosed walls and roof, or the short span.[†]

For such an elegant construct, the design development and preconstruction wranglings were difficult to say the least, as Postel recalls:

"Had we ourselves not been the client, it would never have been built. We had to talk for three months with

Olga Popovic Larsen

Case 2-4 Glass bridge connecting two buildings of the architectural practice Kraaijvanger Urbis Architectural Practice based in Rotterdam.

---

*This is certainly the first glass bridge ever built in Holland, and is believed to be the first such structure anywhere in the world to be designed as a permanent feature of a building.

†It is interesting that one of the employees, who use the bridge to gain access from one building of the architectural practice to the other, took some days of convincing that it is safe to use the bridge. It is amazing how much the feeling of strength is connected with the solidity of materials. One wonders if this has been one of the barriers to using structural glass more widely.

Case Study 2: Glass structures

Case 2-5 The two designers: engineer and architect standing on the glass bridge.

Olga Popovic Larsen

Case 2-6 Gathering of the employees to demonstrate the strength of the bridge structure.

Jannes Linders

the glass industry. They wanted nothing to do with it at first. There had recently been some major problems with glass in the Netherlands, some serious problems with innovative use of glass, and they were afraid of this project. It took three months talking just to convince them that it was possible. Finally they said yes and we took three months for the detailing and the design. We had very odd meetings in Frankfurt Airport and Eindhoven Airport where all sorts of well-paid German specialists were flown in for discussions about how to proceed.* We had three stages of the design: convincing people, preparing it and then making it work.

"Of course we took all the objections extremely seriously. We had to follow the rules to play the game

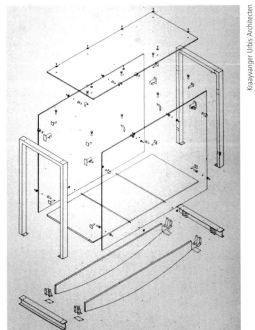

Case 2-7 The glass bridge is a kit of parts.

Kaayvanger Urbis Architecten

*These were in response to a request from the authorities, to ensure the safety margin of the structure.

correctly because we couldn't afford for there to be any problems. As a result, the design development, and discussions with contractors, suppliers, checkers were very detailed, and very expensive. I say the bridge is the most expensive building per m^2 in the Netherlands! About the same cost per m^2 as the Hong Kong & Shanghai Bank, so you can imagine how expensive that one is! And that one is a bit bigger than 4 m^2!"

Expensive perhaps, but worth it if only for the impression which the bridge leaves on the viewer and user. And in fact Postel and Nijsse have used the lessons learned in this work in developing other glass structures. Many other designers also learnt from it. It was the first of many glass structures to follow. Getting consent from the authorities was a lot easier because there had been a successful precedent.

Postel and Nijsse continue to demonstrate their commitment to structural glass in a planned redesign of the reception at the Kraaijvanger Urbis studio, which will incorporate two cylindrical structural glass columns, designed to support the first floor above.

In addition to using Postel's own studio as a showcase, the two have an increasing portfolio of other clients who appreciate the merits of building with glass. Their most recent collaboration is on another glass pavilion in a forest setting in Burgundy, France. Once again they have used glass to minimize the presence of structure and to create a feeling of lightness. The result is beautifully elegant.

The client owned land near the River Serein. On this land there is an old stone and timber folly from the 1700s that had been built as a meeting place for the princess of Orange and her lover. The idea of constructing a new pavilion on the site came with the accidental discovery of a vault in the abutment of a disused

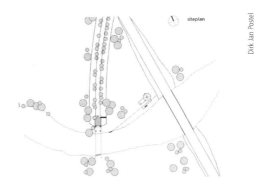

Case 2-8 Site plan of the Temple of Love, wth the old folly to the right.

Case 2-9 View towards the old folly.

Case Study 2: Glass structures

sections

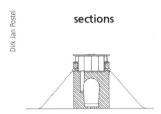

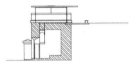

Case 2-10 Section through the new Temple of Love.

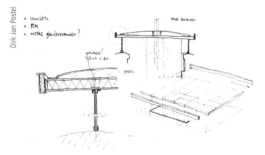

Case 2-11 Sketches showing details of glass and roof connections.

railway bridge. The client who was using the old pavilion as a summer residence decided to construct a small retreat, a modern version of the Temple of Love.

Postel:

"This is a client who likes to build. He likes to have these initiatives. He found the bridge abutment, and on checking it, we discovered a vault inside it. It was an explosion chamber, so that explosives could be packed in and blow up the bridge in times of war. A beautifully made vault, never really made to be seen, except by the engineer with his dynamite! The bridge and the railway were taken away, but the abutment was left. Very well made; big blocks of local limestone. We decided to open up the vault at the side of the river at the base of the abutment. We also wanted to add something over the hatch to complete this idea of a roof over the abutment and to produce an open pavilion. My client said 'I want little structure. I want a view over the whole of the area'. So we thought, well, why don't we make no structure at all? Let the glass do the work. I said I'm sure that it is possible, but I'll call Rob Nijsse."

Technically, the design turned out to be quite straightforward, with one small twist. The roof is made from plywood, and lined with copper sheet, which mirrors the copper used on the old eighteenth-century folly. The vertical load from the roof is carried on the two long glass walls, which sit atop the long stone parapets. This has a double structural advantage of ensuring that the roof spans across the shorter length of the building, and means that the glass walls are relatively short, and so less susceptible to buckling under the compressive load from the roof. The major technical issue was how to laterally brace this load-carrying structure, to make sure that it would not collapse sideways like a house of cards. The standard engineering solution to this problem is to add diagonal bracing in the cross walls. If it is desired to minimize the visual impact of the bracing, a pair of tension cables can be placed in an "X" shape, so that whichever way the building tries to fall, one of the cables will be put into compression and will resist the instability.

Postel:

"Our first idea was to have a panel in the centre of the high glass walls with cross tension bracing. It would be OK, but there would still be *some* structure there. I didn't want that. I wanted nothing at all if possible. So we used the glass panels as shear walls to give lateral stability."

This required the higher walls to be firmly fixed to the roof and base. Now, if the structure tries to fall sideways, shear stresses in the glass provide the same kind of resisting force as the cross-bracing would have done. Shear walls are commonly used in large multi-storey buildings, where reinforced concrete infill panels do the same job, but the use of glass for the purpose is a novel solution.

The result is delightful. From many angles, the glazed walls seem hardly to be there at all. The roof appears to float magically above the forest floor, with the walls providing a shimmering mirage-like reflection of the surrounding forest.

Talking about the design of the pavilion, and his wider design philosophy, Postel highlights his view of the importance of the *architect* appreciating the structural possibilities and difficulties which a project may throw up:

"As an architect, you need to look for the inspiration that your designs will work. Of course I'm not saying that this replaces the engineer; I need help because I cannot do the calculations."

By this, he is emphasizing the importance of understanding structural *form*, which is so vital at the concept stage of a building design:

"If you *don't* know about structure, what happens? You come up with an idea, you send it to the engineer, you get something back and you put that in your drawings and you don't have an interaction. It is much better if you have intuition about form and structure or about how you put it together. You can think about how to create mass on the outside, or how to create a core. The idea of where the material should be. That kind of knowledge gives you a richer feeling about what you design."

In this Postel emphasizes the importance of the conceptual understanding of structural form and the knowledge about materials. He does not argue against the need for an engineer or for an architect to know in detail about structural engineering, but strongly believes that a good intuitive feeling about how a structure "works" can help the architecture. If this is taken to an extreme one can ask whether the duty of an architect should be to look for the "best" structural form, and should architecture be subservient to this? Postel:

"I would argue that it is important to choose the most *effective* structure. This choice depends on the job. We recently had a simple school building project where we looked for the best possible structure as the driving force for the building design, then the architecture is in the detailing. A steel building needs to be well detailed. For

Case 2-12 Two sides of the Temple of Love are fully glazed.

Case 2-13 A view to the surroundings.

Case Study 2: Glass structures

Case 2-14 The roof appears to be floating magically above the forest floor.

a simple building like this, or an industrial building, you don't conceive a very peculiar building. You know it won't be accepted anyway. On the other hand sometimes yes, the building allows you to do something unusual – forces you to do it! And then you do have structural conflicts to resolve and you must work them through with the engineer. To me, there is this fascination of having the structure *help* the architectural design and the architectural design help the structure."

This is a vibrant way of looking at building design: structure and architecture helping each other to create the designer's vision. It is a philosophy which has been put to great use in Postel and Nijsse's glass buildings. It is hardly surprising that the results are so dramatic, and have received wide acclaim. Soon after completion, the beautiful Temple of Love won the Du Pont Benedictus Award in Washington in 2002 for use of glass in architecture, attracting comments from the judges such as:

"I was struck by the poetic clarity in the new as juxtaposed against the old, dramatically aided by the use of laminated glass as the total supporting structure"

and also:

"...probably not a finer example of use of laminated glass as a total structural element."

Postel and Nijsse's creation shows the simple beauty that can come from technical structural innovation and aesthetic working in harmony.

Case Study 3

A Stacked Landscape: Dutch Expo Pavilion 2000. By MVRDV Architects and ABT Consulting Engineers

The Dutch pavilion at the 2000 Expo in Hanover is truly a building like no other, and has rightly aroused considerable interest. Referred to in architectural journals as the "Dutch towering landscape layer cake", in most simplistic terms, the pavilion is a six-storey structure, 25 by 25 metres on plan. However, this tells only a fraction of the story. Few buildings have caves in the basement, a forest on the fifth floor, or a lake and wind-farm on the roof! The building is a stacked landscape in which each floor is a mini-exhibition in itself, with the structure at least as important in the exhibition as the content.

Expo exhibitions occur every four years. The exhibitions are based in a different country and a different city each time. Expos are an opportunity for both the host country and invited guest nations to showcase the best of their culture, industry and business achievements. Each guest nation is invited to construct a pavilion that acts as the centrepiece of their exhibition. Thus, the sense of friendly international competition is intense. This is reflected in the often fascinating, challenging and controversial buildings and structures erected for the events, such as Santiago Calatrava's amazing Alamillo bridge, and the façade of the Pabellon del

Futuro, the wondrously elegant high-tech retake of Gothic architecture designed by Peter Rice for the 1992 Expo in Seville. Or, looking back, the Crystal Palace constructed for the first World Exposition held in London in 1851, which was the first real example of quick-build modular construction.

The 2000 Dutch pavilion was the brainchild of Jacob van Rijs, partner at the young Rotterdam architectural practice MVRDV and structural engineer Robert Nijsse of Arnhem-based engineers ABT.

The layered building reflects the country's relationship with the landscape, something that is of immense importance to every Dutch person. The Netherlands is a densely populated country and for centuries has fought a battle to reclaim land from the claws of the sea.* This continuing battle has given rise to a very particular relationship between the Dutch people and their topography that is represented in the building. MVRDV's ingenious design, which in effect is a monumental, multilevel park, symbolizes the forces of nature modified by man. The building explores the viability of stacking landscapes rather than spreading laterally, and in that way addresses the issues such as the increases in population density, the fragility of nature and the quality of human life.

The pavilion can be described as an open structure consisting of several square planes rep-

Case 3-1 Jacob van Rijs, Architect and Partner at the Rotterdam architectural practice MVRDV, Project Architect for the Dutch Expo Pavilion.

Case 3-2 Robert Nijsse, Structural Engineer and Director of ABT Engineering Consultancy based in Arnhem, Netherlands, Project Engineer for the Dutch Expo Pavilion.

---

*There is a Dutch saying: "God made the world, but the Dutch made Holland". This is a reference to the vast area of land that the Dutch have regained from the sea over the past 1000 years.

Case Study 3: Dutch Expo Pavilion 2000

Case 3-3 Front view of Dutch Expo Pavilion.

resenting the various typical Dutch landscapes: sand dunes, vegetation, real forest, ending with a lake and a roof garden at the top floor. The roof also supports the unavoidable wind generators that provide part of the energy for operating the building. Within these layers of landscape the other facilities, such as exhibition hall, shops, ticket office, information stands, etc., are fitted. The building is wrapped with an open staircase, which provides the main vertical communication through the building. It makes the whole journey through the building an experience and also gives the visitors an opportunity to enjoy the views over the whole Expo site from an elevated position. It is not surprising that many architectural critiques describe the Dutch Expo pavilion as a truly striking building, and probably as the most successful of all at the Expo 2000 in Hanover.

The pavilion was paid for through public funds and MVDRV won a national competition for the right to design it. The competition was

Case 3-4 Internal view, showing the plant pots under the forest layer.

judged by a panel of people representing the Government, industry and cultural institutions. Project Architect van Rijs describes how he developed the fundamental architectural concept behind the design:

"Six architects were chosen to make a proposal. We were one of only two who didn't draw anything at this stage. All the others started to draw or make models, we made more like a story line; what was the issue for the Dutch Expo? We knew we had to do something special and exciting, and also quite challenging, and we made a story about the country, about the construction of our country [land reclamation]. We took this as an opportunity to look into the future at how this mental state may continue. Holland is quite a small country in terms of area, with high population density. We made a proposition to make an exposition of the landscape of the country itself."

This was a project with an enlightened client. While the panel had a list of requirements for the building, they were also looking for the designers to deliver a statement, and help set the scene for the Expo:

"They put the ball at the architects' feet; you propose! That was quite interesting in that there was not a very precise programme of requirements, but they thought that the architecture could generate the theme and the composition of the exposition. The design of the building would start three years before the design of the content of the Pavilion. So this was a very different situation than is usually the case, and we responded by taking the opportunity to develop a story line to the judges. On *that* basis, without a detailed design as such, we were selected. It was a very open-minded decision by the client!"

As previously noted, the scheme proposed by MVRDV was based on a stacked series of landscapes: coastal sand dunes, forests, polders; a microcosm of the physical geography of the Netherlands. (van Rijs jokes that his first ideas included entire floors of tulips and a pig farm on the fourth floor.) This man-made world reflects the intense work that has gone into physically creating the land occupied by the

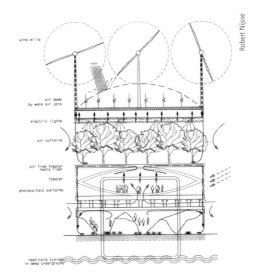

Case 3-5 The stacked landscape and its water balance.

Case 3-6 The stacked landscape and its energy balance.

Netherlands but stimulates debate on issues that will affect building design in the future. As populations increase, can we afford to spread laterally? And if we build our communities vertically, can we avoid the mistakes made in the latter half of the twentieth century, replacing low-rise slums with high-rise ones? In other words, can we provide "natural" environments inside buildings?

At this stage, MVRDV were working alone, without a structural engineer. The same project commission interviewed a number of engineering consultancies for the post, selecting ABT, a company already used to working with MVRDV. Van Rijs says:

"We were quite happy that ABT came out of the decision. We had worked on some small projects from the early days, when our office was just starting up. Actually some of these still get a lot of interest, and keep appearing in publications. We find that we must work as a team in developing the project. The architect plays a role, the engineer plays a role, and it is the *chemistry* that is so important. We knew that we had that chemistry with Rob at ABT."

It is interesting that from the very early stages there was a clear understanding that the final design would be developed through teamwork of the two main designers – the architect and the engineer. And truly, in the development of the final concept, the ideas are bounced to and fro between the architect to the engineer. Van Rijs describes how they started:

"We had the idea of the stacked landscape, and also the idea that every landscape should have its own structure – the most appropriate structure for that landscape. It therefore had a logical concept from that sense, even if the structure appears illogical. Then you pass the ball to the structural engineer and wait for something back."

Perhaps this last comment is a little self-effacing. This is certainly not a situation where the architect has an idea then passes it over to an engineer to "make it work". Both designers speak of how the structural ideas went "backwards and forwards" between the two offices. Nijsse of ABT, who was the Project Engineer for the building, describes the process:

"MVRDV had entered this competition without any advice from a structural engineer. But they had thought about the structure. They already *had* some simple diagrams. Not a full design, but an idea of stacking the landscapes, forests, caves, etc. They had these ideas that they wanted shells, beams columns, etc. in this position or that position. It was very nice. Not realistic, you could not build it like that, but the concept behind it was interesting. But it was really a concept. It was not, start tomorrow and build it. It had very unrealistic sizes. There was a beam on a shell, eh? OK, the image is nice, but it's not very realistic! And of course there are people inside the shell, so not too thick please! But it was a start, an idea. And that for me was very useful, very fruitful. Then I have something to start with. OK what do you want? You want this shell here? You want to hold the building up with it? So how do we make it *work*? What do we need to do to it? That's always the issue; to make it work! And that's the professional challenge."

This is a crucial phase for any building design. In hands less skilled than Nijsse's and van Rijs's, there is a serious danger that the original idea can be lost in the rationalization which the engineer employs to "make it work". Even if the procedure is iterative, rather than sequential, crude imposition of engineering factors can mortally wound the initial architectural concept. Both engineer and architect on this scheme speak of the importance of each of them having clear views and expectations of each other. They both had a clear approach to design, knowing what *they* need to achieve, but also sensitive to the fact that the architecture or the engineering do not exist in a vacuum. Ideally each aspect should work with the other.

Nijsse:

"You start with the concept, but then of course, reality comes in through the door, especially financial reality. As the structural engineer, you must give honest information. The architect asks the questions and you give an honest answer. Sometimes you start to stretch the reality. Perhaps it could work here and you look for the borderline. And for the architect it is the same; if we tell them no we have to put a diagonal in here because it is not strong enough, well then maybe they say, hmmm you really think so? And then of course they must trust me that I am not making an easy way out."

Van Rijs:

"We have a certain attitude that we employ on our projects, and we look for something similar from the engineer. Not to say 'this is how it should be!' and that's it. There should be some potential for changes and dialogue. Maybe it takes a little more time, maybe you cannot do it on every scheme, but on some projects you have the potential for something special, and then you want to invest in it."

This process of constant interaction through dialogue and negotiation happened throughout the whole design process of the pavilion. The structure changed and developed so that the architectural concept remained as true as possible, but at the same time the structure became realistic and one that would work.

Both designers speak of how the initial stage of design included proposals to produce a very simple building frame, with a rectangular column grid running from top to bottom. This would have produced a very inexpensive building, and perhaps freed more resources for the Expo content of the building. In that case more of the funding could have been used for designing the interiors and the exhibitions that

were housed within the building. With a lot of negotiations, the Expo management team were eventually prepared to follow the design team's lead that the building *was* to a great extent, the Expo. The building itself told a story, rather than providing space for a story to be told by the exhibition content. Using this argument the design team managed to manipulate the distribution of funds and reduce the amount for the exhibitions in favour of the structure and the building.

The initial design scheme included a three-dimensional shell at the second floor level, supporting the floors above at its centre, and similarly supported centrally by the floor below. Even though this was not eventually incorporated into the final scheme, Nijsse sees it as an interesting example of the interaction between himself and van Rijs, in particular demonstrating the way in which a provocative initial idea is manipulated into something feasible.

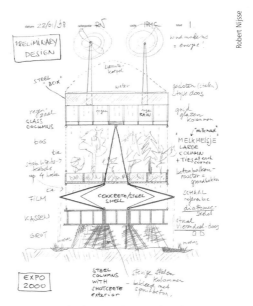

Case 3-7 Design development of structure I-section.

Case Study 3: Dutch Expo Pavilion 2000

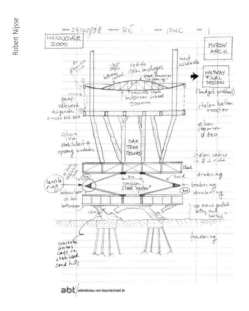

Case 3-8 Design development of structure 2-section.

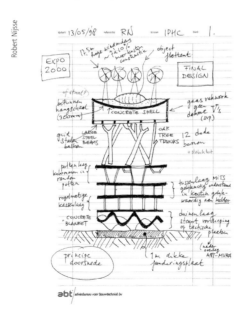

Case 3-9 Design development of structure 3-section.

Clearly, the immediate effect of pushing at the centre of the top and bottom faces of this shell would have been to cause the shells to flatten out, so the engineer's proposal was to encircle the central circumference of the shell with a tension hoop. The next stage was to ensure that the shells were strong enough to transmit the vertical loads in compression from top to bottom. The simple way would be just to make the shells continuous reinforced concrete, but there was an architectural requirement for natural light at this level, meaning that holes would have to be formed in the shells. This pointed the way towards a form that has a precedent in the natural world. Nijsse explains:

"We had to put a building on top of a shell, so we were looking for a very strong form, but also an open one. The diatom is a beautiful example of using little material and still having a very big structural capacity. Beautiful patterns, which of course are also very economical in terms of material use; this combination of properties is the reason they have evolved. So we formulated a similar layout using concrete. In fact, it started to look very much like the slab floors that Nervi made, with ribs following the lines of stress. Little animals worked that one out a long time ago! We wanted to take that lesson and make the structure like that."

In fact, beautiful and elegant as this solution was, it proved to be too expensive to construct. The cost of manpower and materials which would have been required to produce the detailed formwork and arrange the reinforcing bars along the ribs would have far outweighed the material saving. The proposal was shelved in favour of a simple and conventional three-dimensional Vierendeel truss floor.

In a similar way, the development of the forest layer illustrates this iterative and mutually respecting process. The forest layer contains 14 oak columns, and living trees and shrubs, planted in soil in a 1 m deep space under the fifth floor.* Van Rijs's initial plan indicated that he wanted to use the forest structurally. Nijsse had to explain that this would not work.

As discussed earlier, the engineer's role in this situation can be to help the architect achieve a final product as close to their vision as possible, while injecting the necessary reality to point out when the envelope is in danger of being pushed *too far*. Of course, the tips of tree branches have insufficient strength or stiffness to hold the heavy loads from two floors above. A tree in nature has to resist its own self-weight, and the force of the wind. The relative lack of stiffness of the branches is not a major problem; the branches can deflect by certainly many centimetres, and perhaps several metres before they fail. And trees grow vertically. Neither of these are desirable traits for a public building!

Nijsse's initial structural scheme for the forest layer was to support the floor above on a circle of eight columns inset from the building edge. He argued that the columns would be obscured by the trees and would not be noticeable, but van Rijs's ideas were still tending towards the trees. Accepting that living trees couldn't be used as structure, why not use tree trunks in place of steel columns?

In compression, tree trunks are very strong, and their girth provides some degree of resistance to buckling; large tree trunks proved more than adequate to hold up the upper floors, even if *finding* enough 14 m long trunks of sufficiently high quality was to tax the team. But

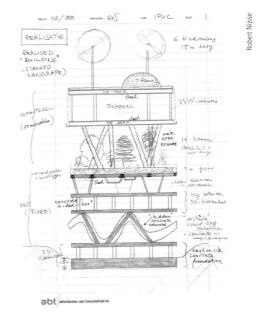

Case 3-10 Final realised design-section.

compressive loads from the vertical weight of a building are not the only forces that columns must resist. Designers must also consider the effect of horizontal forces, either from wind loading or from the natural tendency of a vertically loaded building to sway laterally. In the absence of diagonal bracing, conventional framed structures attain this sway resistance by the columns and beams acting in bending. Engineers use the term *portal* for this rigidly connected frame of columns and beams (reflecting the idea of a laterally stable, yet unbraced doorway). Requiring the trunks to resist bending

---

*The issue of the tree planting shows another aspect of design iteration, and serendipity. MVRDV were initially told that 1 m of soil was too shallow for tree roots, so changed the scheme to include a number of huge plant pots penetrating through the floor into the level below. These are so deep that some of them were extended down to the lower floor level, and finally became incorporated as the structure holding up the fifth floor. In the meanwhile, they found a collaborator who showed them how to make trees grow in only 1 m of soil, similar to bonsai trees, and they reverted to this idea. But by this time, the architectural space created on the fourth floor was integral to the interior design of the pavilion, and the Expo designers were reluctant to have the pots removed. In fact, the final scheme had a light show projecting shadow images of roots onto the (empty) pots. The architect is understandably left with mixed feelings about the result; a wonderful space but perhaps a little deceitful.

Case Study 3: Dutch Expo Pavilion 2000

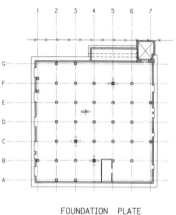

FOUNDATION PLATE

- ▪ concrete column ⌀650
- ◉ concrete column ⌀650+steel tube ⌀559x40
- HD400x551
- (0.00) foundation plate th=1300mm

Case 3-11 Final design – foundations plan.

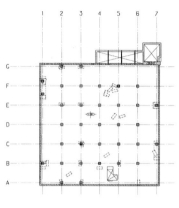

GROUND FLOOR (1st floor)

- concrete column ⌀650
- ◉ concrete column ⌀650+steel tube ⌀559x40
- column disks 1500x500
- (1.00) concrete floor cast on site th=650mm

Case 3-12 Final design – ground floor plan.

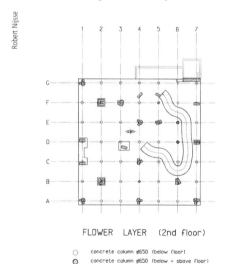

FLOWER LAYER (2nd floor)

- ○ concrete column ⌀650 (below floor)
- ⊘ concrete column ⌀650 (below + above floor)
- ○ steel tube ⌀559x40 (below floor)
- O steel tube ⌀559x40 (below + above floor)
- column disks 1500x500
- (2.00) concrete floor cast on site th=620mm

Case 3-13 Final design – first floor plan.

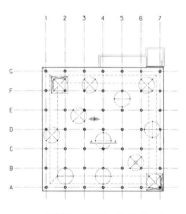

ROOTS LAYER (3rd floor)

- ⊘ concrete column ⌀650
- ○ steel tube ⌀220x220x16
- O steel tube ⌀559x40
- (3.00) concrete floor cast on site th=500mm

Case 3-14 Final design – second floor plan.

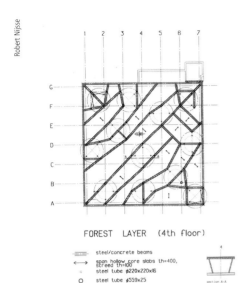

FOREST LAYER (4th floor)

steel/concrete beams
span hollow core slabs th=400, screed th=100
steel tube ⌀220x220x16
steel tube ⌀559x25
hollow core slabs th=400

Case 3-15 Final design – third floor plan.

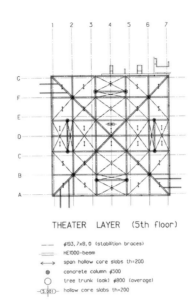

THEATER LAYER (5th floor)

⌀193,7x8,0 (stabilition braces)
HE1000-beam
span hollow core slabs th=200
concrete column ⌀500
tree trunk (oak) ⌀800 (average)
hollow core slabs th=200

Case 3-16 Final design – fourth floor plan.

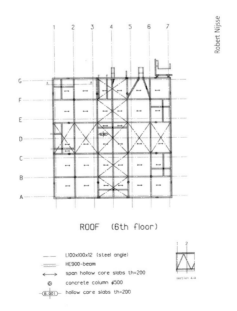

ROOF (6th floor)

L100x100x12 (steel angle)
HE900-beam
span hollow core slabs th=200
concrete column ⌀500
hollow core slabs th=200

Case 3-17 Final design – top floor plan.

stresses as well as the vertical load was a step too far, even without considering how the tree trunks could effectively be rigidly clamped to the floor beams. The possibility of introducing diagonal bracing was a non-starter; steel bracing members are not something one would expect to see in a forest.

The solution was an effective compromise. Although van Rijs's preference was for the supporting trunks to be vertical (since most tree trunks in nature are vertical), the only viable engineering solution was to incline the trunks, and provide the lateral stability by having the trunks act as angled struts. Careful arrangement of the angled trunks could be used to ensure that stability was achieved in all directions.

Nijsse on the ensuing compromise:

"Jacob wanted to have the trees as vertical as possible! And that was a nice discussion! I remember that! Put this one this way…no, no, no, this one this way!"

Van Rijs:

"Ideally we wanted the trunks vertical. Just like a forest yes! But as it turned out it had a nice side effect that some of the trees looked almost as though they were falling. When you looked at the plan there is a worry that there is too much regularity because the structural tree trunks were more or less in a circle; we wanted a forest, so we wanted to avoid this regularity as much as possible. So the diagonals in different directions help to offset this regularity. And in the final scheme, with the real trees around it, you can hardly see any regularity."

The final outcome gives the impression that one is truly in a forest on the fifth floor. At eye level the structural columns appear as real trees (see Figures Cases 3-3 and 3-10). A compromise certainly, but without doubt a successful one.

The design team take a special pride in having pushed the boundary of what is considered feasible. Van Rijs says:

"To make this forest in the air was the greatest achievement. A purpose of the Expo is to make people think, to provoke thought. Now the average Dutchman knows that this is more possible than he thought. Now other people are proposing to put trees in the air!"

For the architect, provoking thought is the issue. For the engineer, the technical challenge provides the excitement, or as Nijsse states:

"We call it stretching technology. A little bit more; a little different. It is very rewarding. Because we have, as structural engineers, a large responsibility; we have to ensure that our structures are safe. Sometimes we are slowed down by that. We say, 'Hold on, we cannot think of every possibility that can go wrong here. Let's not take this risk.' And of course the contractors say the same. They don't want to make strange structures just for the sake of it, because there is a financial risk in doing the unknown. And of course there has to be a balance. Sometimes you have to put your feet on the ground and say, 'No, we shouldn't take the risk; don't do it.' Maybe the architect or the client says, 'I want to make a glass canopy of 20 m span! Can you do it?' And you think, well, this is a big step, let's maybe talk about 5 m first! But yes, sometimes we have to push the border, stretch the way that things go."

Sometimes the desire to innovate is also tempered by the understandable conservative approach of the national or local government checking agencies. In this case, the German authorities, being concerned about the idea of using 14 m long tree trunks to hold up the building, each carrying 200 tonnes of load, demanded the most rigorous tests on the quality of the trees. For the structure to be built, trees that were big enough and also strong enough had to be found. Special tests were carried out in forests by drilling very fine holes to determine whether the tree trunks had sufficient strength to be used on the Expo building. It is interesting that in Holland there were only two oak trees that passed the test. Thus the remaining 12 trees had to be delivered from a Danish forest!

One can agree that pushing the boundaries is anything but easy. However, now that there is a precedent of having a real forest on the fifth floor, we all know that it is possible. It is fair to say that this advancement in knowledge has been achieved through the teamwork of the architect – who had the initial idea, and the engineer – who found a way to make it work.

This real teamwork and the iterative nature of the design process is something that we can find on the rest of the building. For example, the caves level again shows how structural ideas subtly changed the appearance, but also how the practicalities of construction almost led to an entirely different concept.

Van Rijs's original idea for the ground floor of the structure was to have an undulating concrete shell, gently curved to reflect the idea of sand dunes on the long Dutch coastline, but with spaces inside for visitors to walk around, and for usable spaces for the pavilion, for a cafe, toilets, etc. The problem with this, from a structural point of view, is that curved shells are ideal for carrying their own self-weight (see Isler's structures in Chapter 6). A graceful curved shell is not particularly suitable for carrying the very large point loads that would be generated at the points where the shells support the building above. Point loads on a curved shell tend to create very high bending stresses. Thus, the ideal structural shape for a shell carrying point loads is a much straighter line from the position of application of the load to the point where the shell is supported (compare Brunellechi's shape of the dome of Florence Cathedral in Figure 4-18).

Nijsse explains:

"Basically, we started with an ideal arch profile, a parabola, and then, because we needed to make the room under it bigger, we flattened the arches, extending the span. This is a natural process; this is the way you work with an architect – compromising between the ideal structure and the architectural needs. The architect says, we need this amount of space, 8 m, 12 m, whatever and you say, 'OK, the span increases'; you try to deal with the structural consequences. You *react* together. But the structure serves the *purpose* of the building."

It is interesting how Nijsse explains the role of the structure as one to complement the purpose of the building, not compromise it. There are obviously many technical solutions of how to make a building stand up. Some are more efficient than others. However, the one to be chosen *must* always be the one that will complement the purpose of the building and not compromise the design philosophy of the building. In that sense it is important, as in the case of the Expo pavilion, to have true teamwork between the architect and the engineer and dialogue that, through negotiation and iteration, will help in developing an appropriate structural form, even if "appropriate" is not always the most efficient or cheapest structural solution!

Going back to the caves floor, the analysis of shells pushed to this kind of extreme is only practically possible using finite element computer software, which allows the engineer to produce a computer model of the structural form, the material properties, the loads and supports, and determines the forces, stresses and displacements in the structure. Van Rijs and Nijsse determined an acceptable shape for the shells, one that would provide sufficient space beneath the shells, be structurally feasible and still have the required organic flowing appearance, then the engineer created a 3-D computer model of this shape.

Case Study 3: Dutch Expo Pavilion 2000

Hans Werlemann

Case 3-18 Elevation, showing a glimpse of the concrete shell-like structure — bottom right corner.

Nijsse explains:

"We analyzed the shells' structural response to the weight of the rest of the building on top, and found the forces and bending moments in the shells. Because we were using reinforced concrete, we could manipulate the reinforcement; make it stronger where it needs to be, sometimes without even having to thicken the concrete."

Even so, they realized that the shells would need to be up to 1 m thick in places. This was because of the extremely heavy load from the rest of the building. A thicker section both reduces compressive stresses in the concrete, and, because the deeper section gives a greater distance between the compression concrete and the tension steel, makes the shells stronger against bending.

This still left the question of how to construct the shells. Nijsse and van Rijs's idea was that sand hills could be built and shaped to the correct profile of the underside of the shells, and stabilized with cement grout. The contractor could then fix the shells' reinforcement in place and pour the concrete of the shells over this. However, due to the very tight time-scale for the construction, the extremely serious consequences of over-running the construction programme, and the fact that none of the higher structure could be built until the shells were in place, the contractor was unwilling to face the uncertainty of this novel idea. (Novel? See the description of the Florence Cathedral in Chapter 4.) The designers went back to the drawing board. They threw out the idea of the shells and instead looked for something more structurally simple and less complicated to construct. They decided on a complex web of criss-crossing angled concrete struts. This would still give a similar subterranean feel to the discarded shells, but would be much more conventional structurally.

Van Rijs:

"The inclined struts actually made for quite a fascinating concept, even though they were totally different from our original idea, and the storyline of landscapes. But by this time, there were people working on the exposition, and everyone was saying 'There must be a dune layer, you know, we have sand dunes in Holland'."

So dunes there had to be! The simple way to provide the impression of a curved surface while retaining the structural and constructional simplicity of the struts was to construct the structural strut first (thus allowing the construction of the rest of the building to continue), then to wrap wire mesh around the struts and spray a non-structural concrete façade onto the mesh. The architect, still persuaded of the value of his original proposal, is not delighted with the outcome:

"It was not a pure structural expression. It was a fake structure, which in our way of thinking is the worst thing that could happen. Behind it is the *real* structure, but you cannot see it. We thought, these are not real dunes in the way that we had in mind, but we had no option. The dunes had to stay. But in the end, I guess it *looks* OK."

This is an interesting point on which to conclude this case study. It is clear that a continuous process of negotiation, iteration and refinement was necessary, desirable even, in the development of the structural form of this truly fascinating building. In spite of that, some compromises were still unavoidable. This will always be the case when designers try to expand the boundaries of what is known to be possible. Pushing the limits, asking "why can't we do it *this* way?", setting new precedents, *stretching technology* as Nijsse calls it, must always be balanced with the natural caution of those responsible for the safety and economics of a building. The skilful design team can handle these compromises without destroying the original vision. The application of teamwork and understanding can ensure that the quality of a spectacular design survives even as the design is refined. The Dutch Expo pavilion is concrete proof.

Case Study 4

Capsules of Plant Life from Planet Earth: The Eden Project. By Nicholas Grimshaw Architects and Anthony Hunt Associates as Consulting Engineers

The Eden Project, *"a showcase for global biodiversity and human dependence on plants"* is located in a 15 ha former china clay pit near St Austell in Cornwall. It is the world's biggest greenhouse and it consists of two climate-controlled domes (called "biomes", a subgrouping of biosphere, the ecosystems are referred to as biomes) built up from interlocking partial geodesic spheres clad with almost fully transparent foil cushions. This is a truly enormous structure: the larger biome, which recreates the rainforest of the humid tropics, is 240 m long, 55 m high and 110 m wide: enough space to accommodate the Tower of London. During construction, Eden entered the Guinness Book of Records as having both the largest and the tallest structure of birdcage scaffolding erected anywhere in the world. This project was also unique in that, still unfinished, it attracted more than 3500 visitors a day, a figure that has not been exceeded to date by any other structure during its construction stage. Among its many accolades, (which include the filming of scenes from the latest James Bond film at the site), is a description by Isabel Allen (editor of *Architect's Journal*):

> "It has everything: Amazonian rainforests; eerie moonscapes; and ominous shimmering mega-structures which bubble out of the ground. It already has something of a cult status, with media coverage and word-of-mouth eulogising proliferating to such a degree that any formal marketing spend has been kept to a minimum."

Case 4-1 David Kirkland, Nicholas Grimshaw Associates Architects, Project Architect on the Eden Project.

Case 4-2 Alan Jones, Chairman, Anthony Hunt Associates, Project Engineer on the Eden Project.

Case 4-3 The Eden Project, panorama.

Case Study 4: The Eden Project

It is interesting that the concept of a twenty-first century glasshouse in the depths of Cornwall was generated by someone entirely outside the construction industry. Tim Smit, an ex-music industry executive and the driving force behind the restoration of the nearby Lost Gardens of Heligan, had the idea of obtaining Millennium Commission funding for the construction of, perhaps, the most ambitious botanical exhibition and research centre the UK has ever seen.

The project started as a vague concept. The client wanted a greenhouse of a colossal scale, but didn't initially have details on size, position, or even a site! What he did have was a need to convince financiers that this mad scheme had some rationale.

Smit needed to have what he described as a "world-class" design team to provide the practical reassurance that his ambitious idea could be realised. Only in that way did he stand a chance of securing the funding for the project. The team he chose had their most visible advertisement on view for millions of rail passengers entering south London every year. The Waterloo Eurostar Terminal in London, completed in 1993, was the UK's first major mainline railway station terminus for many years. Designed by Nicholas Grimshaw Architects and Engineers Anthony Hunt Associates, it was a suitably bold high-tech piece of architecture. It was to form the basis of the initial design schemes for the Eden Project.

David Kirkland*, a project architect for Eden, describes the client's requirements:

"Tim Smit's initial idea was that he wanted the world's eighth wonder. Lots of plants, had to be big. He want-

ed something that the world would see and wonder at; help to get his message on biodiversity across. He didn't give a brief that the structure should be like Waterloo; the fact that we came up with that was incidental because it fitted the idea."

Smit was entirely upfront with his design team. He didn't have a site, he didn't have the funds for the construction. He didn't even have money to pay for the initial design. What he had was a wonderful vision and an infectious sense that something very special was possible.

Alan Jones[†], Project Director for Eden, recalls:

"This was a unique situation. We were asked to produce a scheme very quickly, very little work. The client didn't have a definite idea of what building he wanted, other than an idea for a big greenhouse. He knew he was asking us to undertake a lot of work for no upfront payment! So we approached this one with a philosophy of how do we produce a solution which has credibility, but requires very little initial work? Initially, we didn't even have a site!"

Or as Kirkland describes the whole situation:

"The economics of this scheme were almost like someone going to a BMW garage and saying, I have £5000, can I have a 3-series? You'd laugh at them. And at first we did, but he, being Tim Smit, said 'Are you saying you're not good enough to rise to the challenge?' So in the end he's got a BMW 3-series for about £10,000, which is not a bad price! He got a very high performance building."

How do you design a building, any building, let alone one intended to be one of the motifs of the millennium celebrations, when you don't even have a site? Hunt's and Grimshaw's brainstormed around the problem. Jones:

Case 4-4 Initial design scheme for Eden Project biomes, model of Waterloo arch-type structure.

*David Kirkland has been part of the architectural team at Nicholas Grimshaw Architects who designed both the Waterloo Eurostar Terminal in London and the Eden Project.

†Alan Jones is Chairman of Anthony Hunt Associates (AHA) and was responsible for the structural design of both the Waterloo Eurostar Terminal in London and the Eden Project.

"Some potential sites were being suggested. One was a hilltop site; the one we finally got was a pit. There was a whole range in between. At one stage we were looking at cable-nets on top of a hill, a huge clear span net. Grimshaw's were very eager to use a cable net because they saw it as a very lightweight solution to a long span roof. For a while, we followed this route; masts, cable nets – very organic shapes."

It wasn't until a likely site, a china clay quarry near St Austell in Cornwall, was identified, that a definitive design could be developed.

Kirkland describes the initial scheme development:

"Normally, to get to a reasonably accurate concept scheme, with good cost estimates, you are looking at spending a significant proportion of the overall design fee. Here we couldn't because the client didn't know if he would ever have the money. We had a classic chicken and egg situation. No money without a design, but no money for the design. We wanted something viable, something with a 'wow' factor, but not to spend too much money on it. A Waterloo-type scheme fulfilled all these requirements, and since we had just finished Waterloo, it was the obvious thing to go for. The original scheme was costed at £100 m, and would be a lean-to against the quarry wall, about 1 mile long.

Jones, on the same theme:

"Once we actually had the site, we very quickly realized that the Waterloo roof which had originally got us the job was actually very suitable for this project. So we settled on a scheme which was geometrically very similar to Waterloo. We had this pit, a cliff-face which we had to enclose. We realized that we could take a Waterloo-type snake and run it along the face of the cliff. It would fulfil the requirements quite well. It would span from the side of the cliff face to the bottom of the pit, over the angle, to give us a covered tunnel leant against the cliff-face. Once we'd done that, the form was generated quite quickly in terms of a series of different sized trusses at as regular spacing as we could get."

Kirkland describes how at this stage the design team came up with the initial design:

"We arranged the initial layout of the greenhouses based on sunlight into the site. For efficient use of solar radiation, you want a wall at the north side rather than glass. The ideal situation for a greenhouse, the traditional layout is a south-facing wall on the north side of the building with a glass front on the south side – a kind of lean-to building. At Eden, we could use the north face of the pit as the back wall, and simply put a glazed canopy over that corner of the pit, 60 m high by 100 m wide. This size gives something that is unique about Eden compared to other glasshouses around the world. The glasshouse tradition was started by the Victorians who were avid collectors. They'd go off to far-off lands and collect plants that were of interest and bring them back home; single samples. If plants could survive at similar temperatures, they would just lump them all together. The unique thing about Eden is that it is a study of populations of plants and biodiversity. So Eden is a collection of representative pieces of, say, the Brazilian rainforest. All the plants are related to that section. You need to be able to ensure that the plants can mature, so you need the size. In most greenhouses, you don't have this size, so you lop off the tops when the plants are too big. This is the reason why the scale required was so huge."

Initially, the major difference between the Waterloo roof and the initial Eden scheme was that the requirement for asymmetry in the arches of Waterloo was absent in Eden, while the scale was quite significantly greater at Eden. Jones, recalls:

"At Waterloo we had from 34 m to 48 m clear span. At Eden, we were looking at about 120 m as a maximum across the diagonal. All we had to do was scale up the arch-trusses by a factor of two or so. Also, at

Waterloo the arches are asymmetrical, so you get a lot of bending as well as the compressive arch action. This is why we used trusses to provide the bending resistance. At Eden, there was less asymmetry, so the arch forces predominated. Also, at Eden, there would just be a single arch span, rather than the three pin arch at Waterloo. At Waterloo, there was a point of contraflexure where the tension member goes from the inside to the outside. At Eden, with the single arch, we didn't have that asymmetry, so we needed a tension member on the inside to stabilize the arch under vertical snow and imposed loading, and a tension member on the outside to stop the arch bursting outwards under wind suction."

To prove to the financiers that this was buildable, the design team prepared a presentation indicating the scale of the scheme against both Waterloo and one of Hunt's previous award winning projects. Jones:

"Although they would be of a different form, the trusses required for Eden were about the same size as those that we had used at the McAlpine Stadium which were 135 m clear span. One reason the scheme was so successful in getting funding was that we were able to go to the funders and say 'If you look at the proposal carefully, it looks very much like Waterloo. It uses the same principles. We've done the scale at Huddersfield, and the geometry at Waterloo, all we've got to do is combine the two. We know what we're doing, we've got a viable solution, we're fairly confident that although this is a big project, it can be done within reasonable cost parameters because we have this experience'."

This really helped in getting the sponsors on board. They understood that the practical side had been considered. The design team demonstrated that this colossal greenhouse could be built.

But, before continuing with the design development of the Eden Project, it is worth looking at the design of the Waterloo Eurostar terminal in London because its structural form was so instrumental in getting the Eden Project off the ground.

The Waterloo terminal was the product of an extremely tight brief. The new station was to wrap its way around dense urban areas of London and the existing Waterloo station. The roof which Grimshaw's and Hunt's produced for the station followed a sinuous curve, constantly changing its span as the width of the station varied in response to the available space. An arch structure was decided on because it requires no intermediate supports and creates a feeling of lightness.

The exact form of the arch was very much governed by site constraints: it had to be almost vertical on one side to give enough clear height for the trains, and on the other side, mainly because of planning restrictions, the roof had to be considerably lower. Thus, the arch was asymmetrical, with the crown skewed over to one side.

Structurally, the effect of making the arch asymmetrical* was to introduce significant bending stresses in addition to the compressive arching action. Looking at the Waterloo profile, it is obvious that the shorter side of the arch is hogging, while the longer section would naturally sag down. These effects require that the arches be made of trusses to give additional bending resistance.

The main arch compression member can be seen to curve smoothly through the span. On the downward sagging side of the roof, the arch is reinforced by placing a tension cable *below*

---

*In any asymmetrical arch, one side will tend to sag downwards causing compression on the top face and tension on the bottom face, while the other side will try to burst (or in structural terms "hog") upwards, with a reversal of compression and tension forces.

this arch and trussing the two together. On the other side, to prevent the arch bursting upwards, a tension cable is laced *over* the arch.

The designers expressed the structural difference between the two sides of the span by tapering each section of the trusses and placing a very obvious pin-joint at the point of contraflexure, where there is no bending moment. Lateral stability of the arch members is achieved by triangulating the truss cross-section. The roof cladding is kept in the plane of the main compression arch member. This ensures a relatively smooth curve for the cladding and also provides significant lateral bracing to the heavily loaded compression member of the arch trusses.

The Waterloo terminal design is a wonderfully expressive structural form and a beautiful piece of architecture. The designers in a very creative way have been able to respond to all the design constraints and especially to the challenges of addressing a complex site.

Returning to Eden, once the initial finances for the Eden Project with a "Waterloo-like" roof design had been secured, design work could start. At the same time the client entered negotiations to buy the site. A major obstacle then arose. The way that the finance was organized meant that the client didn't initially have the money to buy the china clay pit outright. He'd secured an option, but before he could raise the money to buy the pit, the existing owner, knowing that he would soon sell his quarry, stepped up operations. Every time the design team tried to survey the site to fix the geometry of the pit, they would find that the levels had changed, often dramatically. Jones recalls that the very act of surveying the pit exacerbated the problem:

"We were conducting geotechnical investigations as part of the initial survey, checking the ground where we expected to place our foundations in the pit. The

information we were finding was available to the current owner, so every time we found a deposit of china clay, he went in and dug it out! As a result, some places that we wanted to put foundations ended up 10 m below the level that we had originally anticipated."

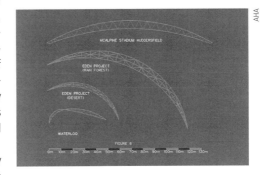

Case 4-5 Initial design scheme for Eden Project biomes, comparison of structure, scale and geometry with previous Hunt projects.

Case 4-6 Waterloo Eurostar Terminal, London, asymmetric arches, hogging section on right, sagging section on left.

Case 4-7 The site, an ever-changing geometry.

This caused major problems. The uncertainty over the actual spans of the trusses meant that detailed design work could not be completed until the site was finally secured.* In the case of Eden, unfortunately, without a fixed span for the trusses, not even the foundations could be finalized, since the foundation loads would be severely affected by changes in roof span. Additionally, the very tight project timetable meant that a delay in the final design of the roof and foundations was simply unacceptable. Jones says:

"We quickly realized that a Waterloo-type solution – individual trusses – wasn't going to work. Every time we designed a truss, it would be a different size when we came to build it. We would have to redesign it all. It was no good saying we will design one truss to span 100 m, one to span 110 m and we will use them wherever they happen to fit, because that would have given us a random-shaped building, which wouldn't have been what either the client or the architect wanted."

Kirkland concurs:

"Principally, we were concerned about the architecture. If we'd just allowed the structure to follow the actual line of the land, it would have been very messy. You have no control over the form. From a technical point of view it would have been buildable. You would have had a lot of problems with twisting, with very different spans close together, but the technology is there to handle that, especially using the foil, rather than glass. At Waterloo we had similar problems, and we had to devise careful details to prevent the twisting causing problems with the glass. With the foil, you can prepattern a twist into it. You could have got round it, but you wouldn't have the form that you had designed."

Another possible solution to the ever-changing ground level, might have been to arbitrarily fix the required foundation levels and then fill in any depressions on site with suitable soil. However, this would have clashed with the philosophy of the design where environmental issues were extremely important. Thus, the material brought onto the site from elsewhere had to be minimized. In addition, the engineers couldn't be certain that sufficient soil of a high enough quality could be located in the site itself to fill the enormous volume of the expected depressions.

At this stage, the design team started discussions with likely contractors about how they would fabricate and erect the greenhouses.[†] Contractors commented that the lack of repetition in the proposed scheme, especially if the trusses were likely to vary significantly in span and size across the roof, meant that the structure would be relatively expensive. They preferred more repetitive ideas; it is clearly much more efficient to be able to fabricate and construct similar basic units many times than to design each item specifically.[‡] These discussions led to abandoning the Waterloo-type scheme. The design team were faced with a radical rethink of their scheme.

---

*It is not unusual for the detailed design of a building superstructure to be worked on as construction of the groundworks continues. In such "fast-track" projects, the designers concentrate on giving the building contractors enough finished design to start construction, then pass on further detailed design data as and when it is necessary. This "just-in-time" management approach is intended to compress the entire design/construction process compared to the more conventional arrangement whereby the entire design is completed before any work starts on site.

†The involvement of contractors early in the design stage of large projects is essential if the final project is to be economically buildable. Without their involvement there is the danger of settling on a design that does not easily lend itself to efficient construction, and will lead to a significant, and unnecessary, overspend.

‡This "off the peg" method of construction and fabrication as opposed to a "made to measure" approach is found throughout the building world. Bricks, timbers, steel beams and columns, all are significantly cheaper if standard and repetitive sizes and shapes are used rather than one-offs. Talented designers are able to embrace this repetition within the language of the building and produce elegant designs.

Kirkland describes the challenges that this entailed:

"The change was very late in the day. Too late in reality. The scheme had been agreed, everyone was on board, the funders, contractor, etc. and the scheme had been getting publicity in the press. We had to change the design, and we had to present to a group of people who were not aware of this. Bankers, client's team, etc. They thought we'd gone mad! But we had no alternative. And very quickly, we had to address the problems of each side, structural engineer, foundations engineer, environmental engineer, client, contractor, etc. We hoped that at the end we would be able to produce an architecturally acceptable scheme."

The team went back to look at their earlier schemes. One of the earliest proposals had been a dome on top of a hillside. The dome had been attractive initially because of its inherent structural efficiency. But at the same time it would have been too symmetrical. Now, with the quarry site, the design team revisited the dome idea, and realized that the site gave them the opportunity both to utilise the efficiency of the sphere, and also to produce a more organic form by combining several domes interlinked together.

Kirkland:

"We had so many objectives to satisfy to make the roofs viable. The client needed a given area and of course wanted to keep the cost down. We studied all sorts of possible solutions, but we kept coming back to a sphere because it maximizes the volume for a given envelope area; this seemed like a good starting point to maximize the efficiency of the scheme. With a geodesic sphere, you can also subdivide it into regular areas, near planar triangles or hexagons and pentagons, which can be very small if necessary, so transportation, fabrication, erection, all these can be optimized. The problem was, we didn't want a spherical building. We'd done a solar model of the site, and had to put the building where the sun was going to

fall. So the building had to be long and against the quarry wall. The way we hit upon to combine these two issues was to join a series of spheres together along the ideal line of the building."

The problem of the undulating ground profile now became an architectural benefit, automatically giving the spherical domes a more organic, less geometrically static feel. The problem of design and fabrication was almost eliminated. Where the spheres intersect with the ground, the foundation line is automatically defined. As the

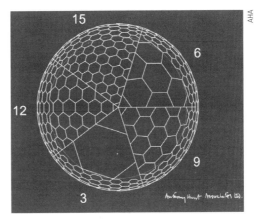

Case 4-8 Hexagon/pentagon geodesic geometry, indicating the effect of different discrete "frequencies" of geodesic layout on panel size.

Case 4-9 Intersecting biome design, original single-layer hexagonal design.

ground level was changing, the spheres were going to intersect the ground wherever they happened to intersect. But because the spheres were made up of discrete members, it simply was a case of adding or removing members to suit the actual final foundation line and level. An additional benefit was that the geodesic sphere is made up of a limited number of discrete lengths of members and connections. Fabrication of these would be considerably simpler than the varying trussed arches.

More changes were to be introduced to the detailed form of the domes: the dome was to be formed from hexagons and pentagons, rather than the basic triangular geodesic dome form. This was for two reasons: to minimize the number of structural framing members to both increase the amount of incoming sunlight and minimize erection time, and to produce near-circular cladding panels, which would lead to a considerably more efficient use of the cladding material foil compared to triangular panels. However, the lack of complete triangulation in a hexagon–pentagon geodesic dome meant that it was not self-stabilizing, as a series of triangles would have been.

To stabilize the domes three solutions were possible. The first option was to use a double layer dome, with a light secondary grid of members added, above or below and attached to the main layer. This would give bending resistance in the way that a truss gets bending strength from separated members. Alternatively, a single layer dome was also a viable option. This could be triangulated by the addition of light tension members across each panel. The third option was to use a single layer grid with larger mem-bers, with connections at the nodes made rigid to provide bending resistance. Jones:

"We believed that the fewer connections we had, the cheaper the solution would be.* So in our design solution, we came up with a single layer dome structure. This relied on fixed connections between the members, and relatively large 500 mm diameter tubes to provide bending resistance for cases where the domes are loaded asymmetrically. This had a minimal number of connections, was very simple, and would still let lots of light through. The architect was very much on board with this. He didn't want a double layer structure.

"We tried to create a braced single layer structure, which would have been very efficient. But the problem with this was that we ended up with bracings inside the cladding panels, because they had to be in the surface of the domes. Because the cladding panels were 2 m deep cushions, you couldn't move this in-plane bracing out of the way of the cladding panels. So, we even went so far as to look at encapsulating the bracing inside the panels. The problem then is that you have a difficulty sealing them where the cushions meet the bracing members at the edges of the panels. Added to that, the cladding contractors didn't like the idea of cushions on the inside of the structure; how could they erect them? Losing the bracing meant that we had to have fixed moment connections. In the end, as a result, the single layer structure was quite a bit heavier than we would have initially liked. There was lots of bending in the elements."

The scheme was sent out to contractors to cost. With the tender results, one of the contractors had offered a revised design, based on a two-layer structure, at a much lower price. They were able to do this because they weren't worried about keeping the number of connec-

---

*Adding a few more tonnes of steel is still relatively cheap if it allows the designer to simplify the connections and fabrication details, which is where the major costs often arise.

tions down. The contractor intended to use a proprietary connection system, produced by Mero Structures.* With the revised design, there were 2.5 times as many members in the frame, but the overall cost was lower because they were simple and easy to fabricate members. By using a double layer frame, the need for bending resistance in the members was reduced, so the overall weight of the roof was reduced by something like 50%. The tube sizes went down from 500 mm to 200 mm diameter.

Interestingly, this design change has made almost no difference to the external appearance of the domes because externally one reads the domes as single layered. The double layer becomes obvious only once you have entered the building. All these reasons were convincing and the design team accepted this radically changed detailed design solution.

Another example of the architect and engineer searching together for an appropriate solution is for the domes' intersections. Where the domes intersect, there had to be quite substantial stiffening arches. These are needed primarily to support the vertical component of force in the

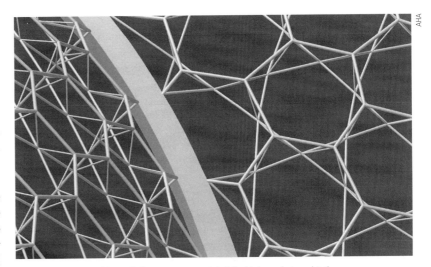
Case 4-11 Intersecting biome design, computer model of double-layer design, detail.

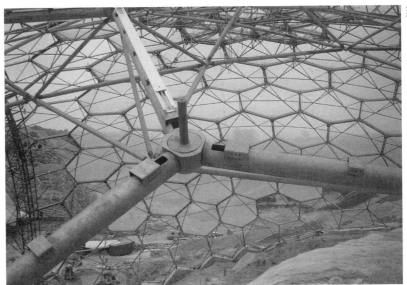
Case 4-12 Construction of the biomes, the Mero system.

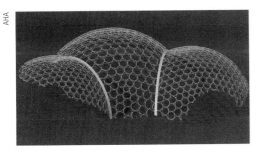
Case 4-10 Intersecting biome design, computer model of double-layer design, general arrangement.

*This is a German company who have been developing and using their own structural components and connection system for frames for over 40 years. This great experience meant that they could estimate with great accuracy what the fabrication and erection costs would be.

AHA

Case 4-13 Mero system connectors. Note the section of a stiffening arch in the background.

AHA

Case 4-14 Standardization of components = economy!

domes. However, as the domes are different sizes, where they meet they also have different *horizontal* thrusts; in other words two intersecting domes would not stabilize each other. The arches therefore also act to distribute these horizontal thrusts back to the supports.

The differences between the domes at the intersections caused the architect even more headaches. Not only are the domes different sizes, but they are positioned differently on plan, and also have different frequencies* within the geodesic format. The result is that the sizes, ori-

entations and positions of the panels of each dome are different. This becomes particularly obvious where the domes meet. The individual members do not meet up in any obviously rational way, and the panels themselves are cut wherever the intersection line happens to pass through them. The different geometries coming in from either side are not *resolved*. Kirkland:

"As architects, we have *visual* issues. I have to confess to being brought up in an architectural culture where all geometry is expected to be resolved. It was very uncomfortable to think that you can have a structural situation where the geometry does not resolve itself.[†] Looking at this and trying different arrangements took about three months. It was a very painful

*Geodesic geometry does not allow the designer infinite flexibility to choose the sizes of the component hexagons and pentagons. They must tie in to mathematical formulae which define how the sphere's surface will be covered by the panels. The sizes which fit these formulae increase in discrete steps, defined by the "frequency" of the geodesic pattern.

†This means that the struts of the geodesic dome do not meet in any obviously regular pattern at the intersections. Also, most panels in the biomes are standard sizes and shapes, but at the edges they are simply cut wherever the intersection line happens to be, in a manner which appears random, and which may seem a little messy. The designers tried different sizes of panels in the geodesic system (which would have meant different overall dome sizes) and tried changing the positions of the intersecting planes, but could not find a solution that resolved this geometry.

process. We thought the detail of the geodesic lines meeting the intersections wasn't good architecturally. We tried to think of ways in which this could be resolved in a considered manner without just letting it be just anything. We began looking at some of these natural concepts such as dragonfly wings. Through balance of pressures, the cells are trying to form hexagons to minimize surface area. When those hexagons intersect, what happens is (and I presume that this is due to pressure), they become perpendicular."

In fact there are very good reasons why these natural forms take up mutually perpendicular arrangements. Mathematical analysis of structural form has considered optimal ways of arranging members in order to maximize the strength of the arrangement for minimum use of material. In structural frameworks, it can be shown that these so-called "optimal" structures often consist of members arranged at right-angles to each other. In the dragonfly wings, and in Frei Otto's soap bubble analyses, Nature has automatically responded to its environment by finding an efficient way of ordering the limited materials at its disposal. Kirkland:

"An architect will look at a dragonfly wing and say 'that is a beautiful thing'. Then you delve deeper, and the geometry is not really resolving itself as we might wish for architectural appearance. Maybe the lesson for us is that we mustn't constrain ourselves with formal architectural hang-ups. So, even though some architects have looked at Eden and found it a bit uncomfortable, I'm happy that the trusses do what they do at these intersections. One of the things that fascinates me is biomimetics; studying nature and trying to learn the lessons from nature. Nature rarely works in a simple linear form."

The architect is suggesting that there is a deeper rationale behind the structural form, which goes beyond superficial appearance. So at Eden, the designers were looking at Nature's

AHA

Case 4-15 Humid/tropics biome nearing completition. Note the unresolved geometry where the geodesic panels meet the stiffening arch.

response to problems and taking their inspiration from that. To change the arrangement, to make it appear more "comfortable" would go against the rationale of the behaviour of the domes. Nature hasn't done it in a dragonfly wing, so why should the designers of the Eden Project? This is a debate which goes well beyond Eden of course, and which, despite the drive at Eden to standardize the components for efficiency, will continue to be fuelled by the power and flexibility of modern fabrication and detailing systems. Kirkland:

"Previously, this unitised, standardized, resolving architecture has been driven by economics and construction systems. You have to do this for efficiency; no one is going to come on board and be able to cut you several thousand pieces of glass to different shapes and do it efficiently. Now that we've got computer-controlled manufacturing processes to pattern things to whatever

Case Study 4: The Eden Project

Case 4-16 ETFE cushion, trial construction and inflation.

shape you want, you are beginning to be no longer constrained to the same degree. So we are entering a new era where we can look at genuinely producing architectural forms that respond to nature, and still be able to fabricate the structures and elements."

The choice of material for the cladding panels was similarly a response to the requirements of the design. Each hexagonal panel is clad using two lens-shaped cushions, with the flat surfaces of the cushions abutting each other along the plane of the panel and the curved surfaces facing outwards and inwards of the building. The cushions are made from extremely thin sheets of ethylene tetrafluoroethylene (ETFE) foil. This is similar in appearance to polyethylene sheet, and is a very different material to the typical woven tension fabrics which are used on tent-like structures and cladding systems such as the Don Valley Stadium roof. Jones explains why this was chosen:

"We could have used flat sheets of glass, but that would have been very inefficient, because it would have meant a lot more dead weight and there would have been much more steel needed in the roof to support it. And of course, the biggest double glazed glass panel you can easily purchase is 4 m x 2 m (we would have needed double-glazed panels for heat insulation).

This means you would have had smaller cladding panels, and kilometres more steelwork or aluminium framing to support the glass and bring the loads back to the primary frame. All of this would have reduced the light getting into the biome. In addition, glass is less translucent than the ETFE to ultra-violet light, which the plants need. Finally, the lifespan of glass double-glazed sealed units is around 20 years. So after 20 years it would have been like painting the Forth Bridge! Permanent maintenance. And of course, a man cannot possibly manoeuvre a 4 x 2 m glass panel onto the roof for replacement. It would need a very large mobile crane pretty much as a permanent fixture on site for maintenance. By comparison, the ETFE panels, 11 m across and 70 m^2 in area can be rolled up, taken up and reinstalled by hand. And the ETFE lifespan is at least 20 years."

The point about the lightness of the ETFE panels is well made. The designers spent a great deal of time trying to maximize the sizes of the panels, to minimize the amount of supporting structure and allow as much light as possible into the building. In doing this, they had to contend with the constraints of geodesic design, together with more practical considerations about just how far they could expect the ETFE to span. From a distance, with nothing to give an impression of scale, it is difficult to judge just how big the ETFE panels are.

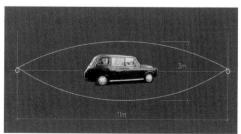

Case 4-17 ETFE cushion, an impression of scale.

The ETFE foil which makes up the cushions is incredibly thin, just 0.2 mm thick on panels spanning 11 m and more. The cushions are inflated by slightly pressurizing the air with which they are filled. The outside surfaces of the cushions then act as tension membranes, curving out of plane and then back to the hexagonal frame. The result is surprisingly strong.

However, there are some downsides to the material. These required careful consideration. The taut foil acts like a drum-skin when impacted by rain. In conventional buildings this could be a significant problem, but at Eden the noise it produces adds to the atmosphere, particularly in the tropics biome. Another issue is that the structural strength of the cushions critically depends on them remaining inflated. Since the foils are so thin, it is possible for them to be damaged accidentally or maliciously, and hence lose their internal pressure. To counter this and to replenish small natural losses in the system, air hoses run along the structural frame and supply air into the cushions from a number of inflation units. The energy consumption is minimal as the pumps only have to maintain a constant pressure in the system, not create an air flow through it. The issue of possible damage to the panels and the extreme ductility of the ETFE led to a particularly unusual design consideration. Jones:

"Before it breaks, the foil stretches to about 400% of its length. This led us to consider the issue of water ponding in the pillows. If one of the horizontal cushions near the top of the domes happens to deflate during a rainstorm, we will start to get water ponding in the panel. Because the foil can stretch so far, the weight of the water could make the panel sag significantly, meaning we could get more water in the panel, etc. We calculated that before the foil would tear, we could get something like 80 tonnes of water

Case 4-18 ETFE cushion, relative ease of construction. (Note the size of the scaffold in the background!)

in the panel. We had to check the structure out to ensure that it could withstand that load if that happened. Also of course, there is the pump system. If you've got a cut in the cushion, all you find is that the pressure drops and the pumps kick in and work a bit harder to refill the cushion and keep it pressurised. It is unlikely that the cushion would deflate completely, and it would take several hours to do so anyway. We have stand-by pumps and a generator, so a complete power failure across the site for several hours in the middle of a storm is the only way that a problem could occur."

An unintended but fitting side-effect of the use of ETFE for the panels is due to its optical qualities. Viewed from outside, the panels are almost opaque, a milky white colour. This adds to the impression of Eden as a self-contained microcosm. But from inside, the panels are

119

almost transparent. Add this to the colossal scale of the biomes, with the roof way above eye level, and after a while the visitor begins to forget that they are inside a building in England. One could truly be in a rainforest. And as Tony Hunt* has said in one of his many public presentations about Eden: *"How many buildings have a 25 m high waterfall inside them?"*

The Eden Project is a unique building in more than one way. It started as a vague concept, even without a site and funding, and has become in the words of one reviewer, the "Capsules of plant life from Planet Earth" housed in the world largest greenhouse. The building has attracted hundreds of thousands of visitors from all over the world and has become a beacon for the area. It is obvious that despite all the challenges, the design team has produced a truly unique building.

Nicholas Grimshaw Architects and Anthony Hunt Associates have a long history of working together. Together they have both worked on real projects and have entered many competitions. Through this they have developed a successful way of working as a team. David Kirkland describes it:

Apex News

Case 4-19 External view of completed humid/tropics biome, a hidden world.

---

*Anthony Hunt is a structural engineer and the founder of Anthony Hunt Associates (AHA) in 1962. He was the Chairman from 1962–2002. He has worked with most of the leading architects of the UK such as Lord Rogers, Zaha Hadid, Sir Nicholas Grimshaw, Will Alsop and others. Tony Hunt is a visiting professor at several UK and American universities, where he teaches structural engineering to architects. He holds an honorary doctorate and is the author of several extremely popular books on structures. These include *Tony Hunt's Structures Notebook* and *Tony Hunt's Sketchbook*.

AHA

Case 4-20 Internal view of completed humid/tropics biome, a rainforest under the sky.

"We have had a very productive dialogue with them [AHA] over the years. I think this comes partly from the fact that, as architects, we at Grimshaw's have a fair understanding of structural principles. Analytically no, but on a conceptual level we have an intuitive understanding of how things can stand up. I guess with Hunt's, this makes the relationship easier. Having spoken to them about this issue, they say that often when they go to an architect's office, the architect describes the structure from an aesthetic angle, then asks the engineer to make it work. That approach has produced some of the worst failings of high-tech architecture over the last few decades. With us, because we have that dialogue, they understand what we are looking for, we understand the constraints of structure. That's something that the companies have maintained over 20 years of working together."

It is important that the team develops an understanding and that through a dialogue of like-minded people the various challenges of the brief are resolved. In the middle ages it was one person who dealt with both the technical and the visual, contextual and aesthetic issues. Today it is a team of like-minded people* who bring into the design their own specialisms and expertise and are able through negotiations to come up with a solution that creatively resolves the design issues. Kirkland describes the mutual understanding with Hunt's:

---

*The team does not consist only of architects and structural engineers. The team will have landscape architects, civil engineers, services engineers, etc. The work of the latter is not addressed in any detail because this book is about bridging the architecture/engineering gap. However, a successful team will have all the necessary disciplines and specialisms represented within the team.

Case Study 4: The Eden Project

Case 4-21 Internal view of completed humid/tropics biome, complete with its own waterfall.

"This understanding doesn't mean mutual compromise. Hunt's for example understand that we appreciate structural concepts. So they are prepared to let us off the leash a little if you like. We have been through many competitions together, which we have not always won, but in which we have tried experimental ideas. Because we have a feel for structural concept, our ideas are rarely stupid. They may be off the wall, but not stupid, unfeasible structurally. Hunt's do their best to make the concepts work. It's much harder for an engineer if the architect doesn't have this feel, this rationale. In these cases, the engineer often has to pull the scheme back, and say 'go down this route'."

Asked about the design philosophy and dealing with technical challenges such as solar gain, site undulation, foundations, etc. Kirkland explains the architect's approach:

"We were driven by pragmatic thinking. I guess an architect's role, with his head in the clouds and his

Case 4-22 The Eden Project, Nature and technology.

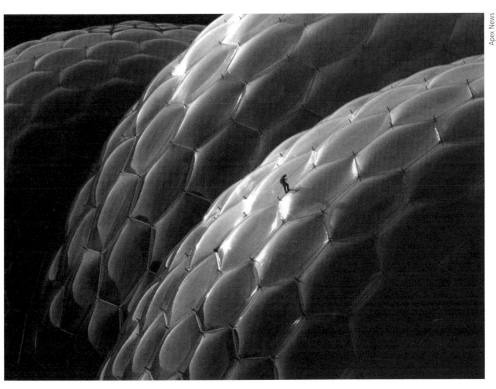

Apex News

Case 4-23 The Eden Project, a synthesis of architecture and structure.

feet on the ground, is to creatively manage objective criteria, and turn them into something beautiful. A lot of people think that we put on a beret and whip up a sketch, then get the engineer to build it. Eden is a wonderful image, but it is a lot richer when people understand how it evolved. It is like looking at Nature, and realizing that Nature's forms are derived from responses to the environment. Eden, hopefully, is the response to its environment."*

Eden is a true response to its environment, a testament to its design team and an architectural landmark for the start of the new millennium.

---

*An interesting postscript to this, and one which reinforces this concept, is being acted out while this book is being written. Grimshaw's and Hunt's are working on the next phase of the Eden Project, and both stress that the design environment for this new project is very different from that which gave birth to the original scheme. Consequently, rather than simply copy the original biomes, they have come up with a concept which responds to its particular driving forces. As we go to press, the detail of this proposal is still kept secret at the client's request, but deciphering the designers' concept should be a fascinating task

Postscript

A successful building must harmoniously synthesize many, often opposing, considerations, of which structural form is only one. Every building, every designer and even every critic will apply different emphasis to different factors. Nevertheless, the success of the design of any building *must* depend to a certain degree on the success of the structural form.

The second chapter of this book contained the following quote from Primo Levi, the chemist turned author who saw aesthetic beauty in technical correctness:

> "In fact it happens in chemistry as in architecture that 'beautiful' edifices, that is, symmetrical and simple, are also the most sturdy: in short, the same thing happens with molecules as with the cupolas of cathedrals or the arches of bridges. And it is also possible that the explanation is neither remote nor metaphysical: to say 'beautiful' is to say 'desirable', and ever since man built he has wanted to build at the smallest expense and in the most durable fashion, and the aesthetic enjoyment he experiences when contemplating his work comes afterwards. Certainly it has not always been this way: there have been centuries in which 'beauty' was identified with adornment, the superimposed, the frills; but it is probable that they were deviant epochs and that true beauty, in which every century recognises itself, is found in the upright stones, ships' hulls, the blade of an axe, the wing of a plane."[1]

Contrast Levi's theory of aesthetics above with John Ruskin's definition of the division between "Architecture" and what he refers to as "Building" (which today would be called structural engineering):

> "That one edifice stands, another floats, and another is suspended on iron springs makes no difference to the art (if so it may be called) of building…building does not become architecture merely by the stability of what it erects…Let us therefore confine the name to that art which, taking up and admitting, as conditions of its working, the necessities and common uses of the building, impresses on its form certain characters, venerable or beautiful, but otherwise unnecessary."[2]

It would have been fascinating to witness a debate on the nature of aesthetics between Levi, the technologist, seeing structure as the leading factor, and John Ruskin, the art historian and critic, viewing structure as superfluous to art. Beauty through technology, or beauty despite technology?

These two standpoints are at diametrically opposed ends of an infinitely graded scale. The position which anyone takes between these limits is a matter of personal opinion.

The twentieth century modernist architect Le Corbusier gives a view of structure (the realm of the engineer) *contributing* to architecture. Unlike Ruskin and Levi, Corbusier's view calls for an acceptance that structure is *one* of the harmonies of architecture:

> "Architecture is a thing of art, a phenomenon of the emotions, lying outside questions of construction and beyond them. The purpose of construction is to make things hold together; of architecture – to move us. Architectural emotion exists when the work rings within us in tune with a universe whose laws we obey, recognize and respect. When certain harmonies have

Postscript

been attained, the work captures us. Architecture is a matter of harmonies, it is a pure creation of the spirit. "The Engineer, inspired by the law of Economy and governed by the mathematical calculation, puts us in accord with universal law. He achieves harmony."[3]

For a building designer from a previous era, a Brunelleschi or a Gothic Master Builder, it would have been difficult to imagine a debate on the contribution of structural form to architecture. For them, structure was an *expression* of their aesthetic vision, without being that vision in entirety. Today, design teams that are able to have a dialogue of equals, understanding each other's design visions and aspirations, can achieve a synthesis which bridges the gap between architecture and engineering.

Notes

1 Primo Levi, (1986) *The Periodic Table* (translated from the Italian by Raymond Rosenthal), Abacus Books, London.
2 John Ruskin, (1906) *The Seven Lamps of Architecture*, George Allan, London, pp. 14–15.
3 Le Corbusier, (1987) 1st edn 1923, *Towards a New Architecture*, Butterworth, Oxford, p. 12.

Bibliography

Architectural Research Quarterly, (2002), vol. 6, no. 3, Clearwater garden: Design research and collaboration, Wigglesworth, S. and Wernick, J., Cambridge University Press.

Baldwin, J., (1996), *Bucky Works: Buckminster Fuller's Ideas for Today*, John Wiley and Sons, New York.

Billington, D. P., (1985), *The Tower and the Bridge: The New Art of Structural Engineering*, 1st edn, Princeton University Press, Princeton NJ.

Blundel Jones, P., (2002), *Modern Architecture Through Case Studies*, Architectural Press, Oxford.

Bowie, T., (1959), *The Sketchbook of Villard de Honnecourt*, Indiana University Press, p. 130.

Brown, A., (2001), *The Engineer's Contribution to Contemporary Architecture: Peter Rise*, Thomas Telford, London.

Chilton, J., (2000), *The Engineer's Contribution to Contemporary Architecture: Heinz Isler*, Thomas Telford, London.

Crossley, F. H., (1951), *Timber Building in England, from Early Times to the End of the Seventeenth Century*, Batsford, London.

Drew, P., (1976), *Frei Otto Form and Structure*, Granada Publishing, London.

Gordon, J. E., (1978), *Structures, or why things don't fall down*, Penguin Books, London.

Happold, E., (1984), The breadth and depth of structural design, in *The Art and Practice of Structural Design – 75th Anniversary International Conference of the Institution of Structural Engineers*, July, pp. 16–22.

Harvey, J., (1971), *The Master Builders – Architecture in the Middle Ages*, Thames and Hudson, London.

Holgate, A., (1986), *The Art in Structural Design*, Clarendon Press, Oxford.

Hunt, T., (1997), *Tony Hunt's Structures Notebook*, Architectural Press, Oxford.

Hunt, T., (1999), *Tony Hunt's Structures Sketchbook*, Architectural Press, Oxford.

King, R, (2001), *Brunelleschi's Dome*, Pimlico Random House, London.

Laugier, M. A., (1977), 1st edn 1755, *An Essay on Architecture*, Hennessay & Ingalls, Los Angeles.

Le Corbusier, (1987), 1st edn 1923, *Towards a New Architecture*, Butterworth, Oxford.

Le Corbusier, (1960), August, *Science et Vie*.

Levi, P., (1986), *The Periodic Table* (translated from the Italian by Raymond Rosenthal), Abacus Books, London.

Levy, M. & Salvadori M., (1994), *Why Buildings Fall Down*, W.W. Norton, London.

Macdonald, A., (2000), *Anthony Hunt: The Engineer's Contribution to Architecture*, Thomas Telford, London.

Mainstone, R., (1975), *Developments in Structural Form*, Allen Lane, London.

Manetti, A., (1970), *The Life of Brunelleschi*, University Press, Pennsylvania.

Marks, Robert W., (1960), *The Dymaxion World of Buckminster Fuller*, Reinhold Publishing Corporation, New York.

McHalle, J., (1962), *R. Buckminster Fuller*, Prentice Hall International, New York.

Mower, D., (1977), *Gaudi*, Oresko Books, London.

Murray, P., (1986), *The Architecture of the Italian Renaissance*, Batsford, London.

Oliver, P., (1987), *Dwellings: The House Across the World*, Phaidon Press, Oxford.

Polano, S., (1996), *Santiago Calatrava: Complete Works*, Electa, Milan.

Bibliography

Reitzel, E., (1972), *Fra brud til form*, Polyteknisk Forlag, Denmark.

Ruskin, J., (1906), *The Seven Lamps of Architecture*, George Allan, London.

Salvadori, M., (1990), *Why Buildings Stand Up, The strength of architecture*, W.W. Norton, London.

Serlio, S., (1970), first publ. 1619, *First book of Architecture*, Benjamin Bloom, New York, p. 57.

Sieden, L. S., (1989), *Buckminster Fuller's Universe. An Appreciation*, Plenum Press, New York & London, pp. 102–103.

Spasov, A., (1994), *Gradezni Konstrukcii*, Univerzitet Kiril I Metodij, Skopje.

The Architectural Review, (2002), Total landscape, MVRDV, September, pp. 65–67.

Index

Page numbers in italics refer to illustrations, The suffix "n" indicates a reference to a footnote.

Index

Index